Human Factors in Performing Arts

This captivating book explores the intersection where performing arts meets human interaction and delves into the application of human factors principles in this field. From music and theatre to cinema and magic shows, indoor and outdoor performances are analysed from a human factors perspective.

Written in an accessible language, this book offers a comprehensive overview of how human factors influence various facets of the performing arts, enriching the experience for both performers and audiences alike. This book uncovers how human factors principles can enhance performance across script writing, stage design, crowd engagement and more. Through engaging storytelling and practical examples, readers will learn about the intricacies of audience attention, ambience creation and the importance of feedback in shaping memorable performances. Whether you're a seasoned professional or an aspiring events manager looking to enhance the user experience, this book provides valuable insights into optimising the human experience within the realm of performing arts, making it safe and memorable for all.

Ideal for professionals in human factors, occupational health and safety, as well as those working in the performing arts industry, *Human Factors in Performing Arts: A Layperson's Approach* serves as a valuable resource for theatre managers, event organisers and anyone involved in orchestrating small- or large-scale performances. With its blend of theory and real-world application, the book offers a fresh perspective on human factors and the art of entertainment, making it essential reading for anyone passionate about elevating their craft.

Human Factors in Performing Arts

A Layperson's Approach

Prabir Mukhopadhyay, MSc, PhD

CRC Press
Taylor & Francis Group
Boca Raton London New York

CRC Press is an imprint of the
Taylor & Francis Group, an **informa** business

First edition published 2025
by CRC Press
2385 NW Executive Center Drive, Suite 320, Boca Raton FL 33431

and by CRC Press
4 Park Square, Milton Park, Abingdon, Oxon, OX14 4RN

CRC Press is an imprint of Taylor & Francis Group, LLC

Library of Congress Cataloging-in-Publication Data
Names: Mukhopadhyay, Prabir, 1969–author
Title: Human factors in performing arts : a layperson's approach / Prabir Mukhopadhyay.
Description: First edition. | Boca Raton : CRC Press, 2024. |
Includes bibliographical references and index. |
Summary: "This captivating book explores the intersection where performing art meets human interaction and delves into the application of human factors' principles in this field. From music to magic shows, performances are analyzed from a human factors perspective. This book offers a comprehensive overview of how human factors influence the nuances of performing arts. Ideal for professionals in human factors, occupational health and safety, and those working in the performing arts industry, Human Factors in Performing Arts: A Layperson's Approach serves as a guide for theatre managers, event organizers, and anyone involved in orchestrating small or large-scale performances"– Provided by publisher.
Identifiers: LCCN 2024010981 | ISBN 9781032826912 (hardback) |
ISBN 9781032828152 (paperback) | ISBN 9781003506409 (ebook)
Subjects: LCSH: Performing arts–Production and direction. | Human engineering.
Classification: LCC PN1590.P74 M85 2024 | DDC 791–dc23/eng/20240326
LC record available at https://lccn.loc.gov/2024010981

ISBN: 9781032826912 (hbk)
ISBN: 9781032828152 (pbk)
ISBN: 9781003506409 (ebk)

DOI: 10.1201/9781003506409

Typeset in Times
by Newgen Publishing UK

Dedication

To the Lotus Feet of Shri Ramakrishna Paramhansha Dev, Pranam to You…

Late Dhirendranath Mukhopadhyay (My Father)

&

Mrs Meena Mukhopadhyay (My Mother)

&

Dada, Mamoni, Duggi and Leto…

&

To those performing artists who bring in joy to the lives of millions of common people, struggling hard to make their both ends meet.

Contents

Preface

In the field of performing arts, which is a big domain starting from music, theatre, cinema, magic, circus, etc., there have been hardly any books that talk about human factors intervention. A few research works that are available in this domain are too technical for the common people to understand. Thus, all those associated with performing arts barely apply human factors principles, the majority of them being from arts, humanities and similar disciplines. The technical and scientific approach of human factors makes them jittery to use the subject in their profession. In this book, human factors and ergonomics mean the same.

This book makes an attempt in that direction. Human factors as a subject has immense potential in all the professions which involve humans and performing arts is one such profession. Any performing arts ranging from magic, opera, circus, theatre, music, etc., which aims at entertaining and enriching the audience, can benefit a lot by applying human factors principles.

The writing is in a storytelling format, with examples from real world, which should make the book interesting to read and at the same time easy to apply in the real world. Each chapter touches upon different aspects of performing arts. They start with an "overview" section that prepares the readers what they can expect out of the chapter and ends with "Key Points" which are a gist of the chapter. This is followed by a "Practice Session" with "human factors directions" which would help the readers test out how much they have understood. The last chapter is dedicated to different types of exercises with human factors directions. The exercises have been carefully selected from different aspects of performing arts and its human factors application.

All the images in this book have been taken from royalty-free sources. They are for representation purpose only and build the storyline. The line diagrams have been created using Google AutoDraw. I hope the readers enjoy reading the book. Any comments and suggestions are always welcome.

Dr Prabir Mukhopadhyay
Email: prabirdr@gmail.com

Author Biography

Dr Prabir Mukhopadhyay is Associate Professor and Head, Design Discipline, Indian Institute of Information Technology Design and Manufacturing Jabalpur, Madhya Pradesh, India. Prabir holds a PhD in industrial ergonomics from the University of Limerick, Ireland. He started working with noted ergonomist Prof. R. N. Sen at the Calcutta University both for his master's thesis and later on a project sponsored by the Ministry of Environment and Forests, Government of India. It was during this time that Prabir developed a keen interest in the subject and wanted to build his career in ergonomics. He joined the National Institute of Design, Ahmedabad, India, as an ergonomist for one of the projects for the Indian Railways. He then joined the same institute as a faculty in ergonomics.

During his tenure at Ahmedabad, he worked on many consultancy projects related to ergonomics. Some of his clients there included the Indian Railways, Self Employed Women's Association and the United Nations Industrial Development Organization.

After working there for two years Prabir left for the University of Limerick, Ireland, on a European Union Funded Project under the supervision of Prof. T. J. Gallwey. He completed his PhD in industrial ergonomics from the same university and decided to return to India to apply his acquired knowledge. He joined the National Institute of Design, Postgraduate Campus at Gandhinagar, India, as a faculty in ergonomics. There he headed the Software and User Interface Design discipline. Simultaneously, he started teaching ergonomics across different design disciplines at other campuses of the institute such as Ahmedabad and Bangalore.

After working there for around five years, Prabir joined his present institute as an assistant professor in the design discipline. He was later promoted to an associate professor and later became the discipline head. He teaches, practises and researches in different areas of ergonomics and its application in design. He has authored six books till date without any collaboration, which include *Ergonomics for the Layman: Application in Design* (CRC Press, 2019), *Ergonomics Principles in Design: An Illustrated Fundamental Approach* (CRC Press, 2022), *Ergonomics in Fashion Design: A Layperson's Approach* (2022), *Visual Ergonomics for Communication Design: A Layperson's Approach* (CRC Press, 2022) and *Application of Ergonomics in Handicraft: A Layperson's Approach* (2023) and *Human Factors in Tourism: A Layperson's Approach* (2024). His favourite pastime is writing, travelling and cooking.

Acknowledgements

The line diagrams have been drawn using Google AutoDraw free online AI tool. We sincerely acknowledge Google AutoDraw. All the images used in this book have been taken from royalty-free sources. We sincerely acknowledge all the photographers for allowing us to use their images for free. The feedback given by all the anonymous reviewers is also deeply acknowledged.

Introduction to Human Factors in Performing Arts and the Dimensional Aspects

1

1.1 OVERVIEW

This chapter gives the readers an overview of the application of ergonomics principles in different facets and types of performing arts. It discusses how the systems perspective of human factors can help in identifying the different weaker links and help to strengthen them so as to minimize error and increase performance. Then the chapter presents different application areas of human factors in performing arts, with a focus on enhancing productivity and augmenting interaction between the artist and the audience. The chapter also provides the readers with an overview of the importance of ergonomics and body dimensions, also called anthropometry, in performing arts. Different dimensional aspects of musical instruments, props, space for performance and the dimensional aspects of the costume worn by the performers are described. The dimensional aspects related to audience seating and standing which lead to performance enjoyment are also discussed here. Ergonomic aspects of dimensions are discussed in different types of settings, both formal and informal, outdoor and indoor.

DOI:10.1201/9781003506409-1

The chapter concludes with some guidelines on accommodating the artist and audience with extreme body dimensions and to ensure that they gel with the others, the so-called "average" body dimensions.

1.2 INTRODUCTION TO HUMAN FACTORS AND PERFORMING ARTS AS A PROFESSION

Mr Leto is the lead singer of a famous band "Bikae Monto". His team comprises members who play the guitar, drum, saxophone, keyboard, and flute. The band performs in different parts of the country and the world. Their performance is appreciated and they were going popular day by day. Once they were invited to perform at a specific venue. The team reached the venue and Leto moved forward to go on to the stage. The staircase to the stage was so small that he could barely keep his feet on them with his sports shoes. He was about to fall when one of the members of the band held him just on time. Finally the band members all went on to the stage to take their respective positions. The band members found that after positioning their musical instruments there was hardly any free space left for Leto to sing and dance and hence make the band much livelier. The stage was covered but the audience were standing in the open field. The performance started with the sound system being checked and all the instruments synchronised for a seamless performance. The lights on the stage were also checked. With an announcement the programme started. The lights went on and the musical extravaganza started with the saxophone. Leto started singing his opening song for the event. Things were going on well, the concert picked up but the band was finding it tough to accommodate on the stage. Leto started his movement and all of a sudden the stage was dark and Leto could not anticipate the end of the stage and fell down. Luckily he was picked up and started once again. Today it appeared that the team lacked coherence. This was the third concert of the day and they were feeling drained out. One of the musicians playing the flute had a problem at home as his wife was admitted to the hospital after meeting with an accident. The performance finally ended but somewhere the audience felt that the band "Bike Monto" could not meet their expectations today. Leto and his team on their return journey in the car started discussing what went wrong and how it could be improved upon in the next performance.

A closer look at the above incident points to some very important issues related to "performance" in an art like this. The band's performance was affected by the dimensions of the stage which was believed to be too small.

Thus, the artist's body dimensions, movement on the stage, and movement and space requirement for playing the instruments needed to be considered before the performance started. This was the band's third performance of the day. Such performances need very high energy level and lead to exhaustion on the part of the performers. If adequate rest is not given after a three-hour performance, the body cannot recover from the minute wear and tear that the tissues undergo. Apart from this, the brain which is the command and control centre involved in processing of information and coordination between different body parts also gets tired. One of the musicians' wife was admitted to the hospital and that affected the musician's performance who was playing the flute. In such cases, if one person goes wrong, then the synchrony of the team goes for a toss. Thus, any performance needs a careful consideration of artists' movement patterns, the amount of space required, their mental state and their level of exertion. When all these things are considered and balanced, one can expect a better performance. This is exactly where human factors can come to the rescue. In a simple language it's the interplay between musicians, the space and the venue, their movement and their mental and physical states affecting performance (Figure 1.1). This is a subject which helps you in looking at people and how to enhance their efficiency and make their life comfortable. This is achieved by prudent use of human capability and taking care of the elements like stage area,

FIGURE 1.1 The macro and the micro picture of human factors in performing arts. Photo by Haste LeArt V: www.pexels.com/photo/six-people-in-black-matching-clothes-dancing-at-daytime-690597/

spacing between musicians, extra space for performers to move, jump, dance, etc. It also considers the heat and ventilation of the space so that performers are comfortable and capable of giving their best. It ensures the illumination level – should the lights fall directly on performers' face (it would distract them), how loud the acoustics should be, etc. So the musical performance or the concert is affected by the people performing: their physical and mental well-being, the dimensional aspects of the stage, the reaction of the crowd and the comfortable environment (neither too hot nor too cold) are factors that need to be considered. A good performer might not perform well if these elements are not factored in. Thus, any performing art demands that we look into the art from a macro and a micro perspective. Sounds little confusing? Let me give you an example. Try to visualise the performance venue, a big field, a stage at the end with a nice back drop, musicians with their instruments positioned on stage and the singer singing. The lights are illuminating the stage and the acoustics are all booming with energy. Along with this there is a huge crowd all charged and excited, dancing to the tune of the band. This harmonious picture that you visualise is called the macro picture. Now just like a camera focus only on the lead singer, his dress, facial expressions, the type of trouser and shoes, etc. This is the micro picture. So in performance like this you need to see both the micro and macro pictures because the micro elements connect with one another to make the macro picture (comprising the performance, stage, audience, props, etc.). This is what the subject ergonomics does; it enhances performance by identifying where should we focus in the micro elements and how to ensure a perfect macro picture to enhance better productivity or performance so that the concert becomes a grand success.

1.3 THE SYSTEMS PERSPECTIVE IN PERFORMING ARTS AND ITS NECESSITY

The word "system/s" is used very frequently in our daily life. You hear your dad complaining how the government officials in the city corporation do not cooperate with the general public and he says "the system is corrupt". You hear your school teacher at times complaining that the system should be working better. Have you thought what exactly do these people mean by "systems"? Why is it necessary to look at problems around you through the lens of systems? We need to address two issues here: what are systems and why it is necessary when we talk about ergonomics and its application. You have all gone to a circus. In any circus there is a very popular performance known as the trapeze. This is where the performers perform different acrobatics high in the air just on a few rods attached to the top of the circus tent. It's such a coordinated effort that when one person leaves his/her hand the other trapeze rod is within his/her reach (Figure 1.2). The interesting part is that each performer has to play his/her role accurately; else if there is one mistake on part of anyone in the group, then the entire trapeze act falls flat and looks something out of gear. This entire trapeze act is an example of a systems concept where two or more performers interact with one another for putting up a great show of acrobatics. To the audience it looks like a single performance of acrobatics, but in reality it's the concerted effort by every performer who contributes to this spectacular performance. This is an example of a system where two or

FIGURE 1.2 Trapeze act in circus as a systems. Photo by Robert Linder on Unspalsh.com.

more entities interact with each other for a common purpose. When we look at performing arts through the lens of systems, the smaller parts, their role, problems, etc., become clear and this paves the way for human factors intervention. In an orchestra only when everyone plays their instruments perfectly that the orchestra is perfect. If one team member plays a little out of tune, then the orchestra goes out of tune even if others are playing perfectly. This is the power of looking through the lens of systems at any performing arts so that we could get to know where improvements can be made and in what manner. Is the problem with users or their instruments/props, or is it with both? These questions can be answered through systems approach.

1.4 WHERE ARE THE LOOPHOLES FOR INTERVENTION?

In any performing arts, there are many vulnerable points which demand a human factors intervention. You are familiar with street performances. It could be a music, orchestra, band or street drama. There are minimal special effects with no stage, light or acoustics. Such performances are very difficult because there are a lot of distractions in the vicinity with people moving around, uncontrolled noise, natural hazards like rain, sun, snow, etc. These performers have to put in extra effort to entertain the audience and thus they have to get their performance tailored in accordance with the context in order to put up the best show (Figure 1.3). If you see this performance through the lens of a system, then the factors mentioned above are the different components which make the system or the macro picture of street performance. So your intervention in terms of ergonomics should try to cover or make the weak points robust so that the system becomes strong and the performance improves. How do you do it? As there is lot of noise around, any musical performance in this context should be loud and should have a lot of pitch variations. The illumination level might be inadequate and to compensate for that performers have to wear clothes of very bright colours so that they are seen from a distance. As there are no stages, it's better if the performers use props, and exhibit a lot of movements in case of street plays to draw attention of the spectators.

Let us come to the case of an opera performance. It's a big auditorium. A controlled environment where there is absolutely no communication with the outside world (Figure 1.4). Here the system comprises the stage along with the audiovisual system, artists, their dresses and props. The other components are the seating for audience, the audience themselves, the staffs and the security. Coupled with this and in order to ensure

FIGURE 1.3 Human factors in street performance. Photo by Airam Dato-on:www.pexels.com/photo/group-of-men-playing-music-15189444/

utmost comfort of the audience, the temperature of the auditorium needs to be comfortable both for the performers and for the audience. Lastly, the most important aspect of the system, safety for all – in terms of the exits, their number and locations – is of paramount importance. If all these components are in place, the performance would be successful. If any one component fails, the quality of the performance suffers. The vulnerable points where one should focus on and ensure ergonomic interventions are public address system and the people associated with them should ensure it is audible for all. The lighting system both for the stage and for the auditorium should be synchronised in a manner so that when the stage lights

FIGURE 1.4 Human factors in an opera performance. Photo by Matheus Viana: www.pexels.com/photo/people-watching-play-on-stage-2372945/

are on the auditorium lights are off and vice versa. If this control is manual, then things can go wrong and hence there should be technology to augment human controlling of this element. It's possible that performers might fall ill due to various reasons and one of them is heat stress. This is where it's essential to maintain comfortable temperature of the stage and the auditorium at large. A staff would be at the control for the same but again an automatic control monitored by a staff would ensure that this part of the auditorium (subsystem in the system) is robust and has fewer chances of failures. You have to remember that in a system for performing arts there is always an input being given and an output being received and both these involve humans. These two points where we give input and receive output are vulnerable to error. Ergonomics has to make these vulnerable points robust by accounting for the prudent usage of technology. This is because people are different and the system is fixed. This is the biggest challenge for ergonomics in increasing systems reliability and ensuring that the show must not only go on but should go on in the same way every day and should go well.

1.5 THE PROFESSION AND THE CHALLENGES INVOLVED

Performing arts is a large domain which comprises different types of performances like music, band, theatre, magic, circus, etc. The setup also has a wide variety, ranging from controlled environment in an auditorium to outdoor environment to street performances as well. Each of them has their own unique characteristics and challenges. Ergonomics as a subject can intervene and augment performance at different levels and in different ways.

Ms Tubri is a vocalist. She is young and a college-going student who prefers to earn her "pocket money" by singing in the local train. She uses her Spanish guitar and sings in the train after her college hours. In the morning the train is not that crowded so she has no problem as such. In the evening after her college is over people start heading homewards and there is heavy crowd in the train. It's here that she faces the problem. It's not only the problem with standing in the space inside the train compartment with her guitar, but also making herself audible to all, because she does not use any public address system. To enhance her performance in such a context ergonomic interventions are necessary. The challenge is that it has to be feasible and cost-effective. Buying a public address system is expensive for her. So she plans to sing on the railway platform at the place where people leave the platform. This way she is able to get a space for performance and is able to draw the attention of people as well. After a tired day of work, the crowd often stand and listen to her music and the earnings are also good. At last Tubri could position herself at a better location based on the public mood and convenience.

Mr Leto is a musician who plays the saxophone. He performs along with his team of musicians comprising a base drum, a keyboard and an octapad. The major problem that Leto faces is that it's difficult to get a musician to play the base drum. So what's happening is that there is a change in the drummer at frequent interval and it takes time for Leto and his team to get the coordination back to track. Many times they do not get enough practice with the new drummer and if they have a performance already scheduled it goes a little off-beat and the performance does not go that well and suffers. This is because in musical performance coordination takes time and with frequent change of the drummer it takes time for the new drummer to adapt to the band.

Mr Dhikchao is a young film director and makes short films. He is still struggling to establish himself in the industry. The problem he is facing is fatigue at the end of the day which is affecting his performance. He has to work for long in awkward posture looking at the camera with almost no rest. This constrained posture is leading to pain in different parts of his body at the end of the day. Thus, he has to take rest for a few days before the pain subsides and then resume his work. Dhikchao is looking for a permanent solution to this problem. His friend has advised him to take micro breaks in between work spells – for example, after every hour of work a minute of break wherein he has been asked to stretch his body parts and bend in the reverse direction. This technique seems to be working for Dhikchao.

All these examples are depictions from the world of performing arts where the problem mainly lies in or is related to "humans" in a specific context. If only you inculcate the habit of seeing the problem through the lens of "humans" along with the context that your solutions would be prudent and would work.

1.6 TYPES OF PERFORMING ARTS AND THE HUMAN FACTORS LINK

There are different types of performing arts and each has its own uniqueness. There are some within a format and a controlled environment. Stage performances fall under such category. There are controlled light and sound arrangement and the performers come on stage after rigorous rehearsal. The audience also are seated in defined seating position. The entire performance goes on like planned by the organiser and ends on time. If you look at this, it appears to be too simple and nothing can go wrong. In this case the primary users are the audience for whom the performance is. They are sitting at one end of a system and absorbing what is being offered to them. At the other end of the system are the performers. The infrastructure and the props used by the performers are other parts of the system or of the entire macro picture which makes up any performances. If one wants to improve performance, what are the steps and how to proceed? The performers and the audience are always at the two distinct ends of the picture. As any performance is for the audience and not for the performers, so the feedback from the audience is important for improving the performance. Research needs to be done on the needs and wants of the audience and what they would like to see and what not. What else affects performance? The health and well-being of the performers. If the performers fall sick either physically and mentally, then they would not be able to give in their best. In addition, the dresses they wear should be comfortable and not lead to movement restrictions or make them feel too hot or too cold. The props that are used should be easy to handle and should be in tandem with the body dimensions of the performers. The acoustic system should be in place so that they don't have to exert their voice beyond their limit. The lights that illuminate the stage should not fall into their eyes, else it would be difficult for them to see and could lead to visual fatigue. There are many factors that can affect performance and ergonomics can ensure that all those factors related to those who perform are given that "humane" touch, thus ensuring a seamless integration with the performers.

In any performance the application of ergonomics to enhance performance happens in the same way as mentioned above. There are different types of performing arts in this world, ranging from cinema, theatre, music, magic, circus, etc., and the list goes on. These are performed in different setups, at times in the open, sometimes in closed environment. At times these also happen in crowded places in the streets, market place, public transports, etc. Wherever it is performed there is a clear division of zones between the audience and the performers. This zone is not always defined, the audience might surround the performers and they might stand in between them as well. Human factors as a subject can be applied in each of these situations to enhance performance and ensure that the audience enjoys. The point of vulnerability in performance is that when the performance is being communicated to the audience and how well it's being done. There should not be any barrier between the performer and the audience and ergonomics has to ensure that. This could be achieved by careful crafting of dialogues, proper acoustics, movement pattern of the performers and ensuring that performers and the audience are positioned in such a way that it could facilitate communication in the best possible manner with least distraction.

1.7 THE ARTIST IN PERFORMING ARTS

The artist is a human and should be energetic and full of confidence. The artist needs feedback from the audience which boosts her/his morale for better performance. This is applicable in live performance and not in recorded one. We have to remember that every artist is different just like any individuals. Thus, always ensure that the artist is comfortable, and should not fatigue while performing. If he/she fatigues, then either the exhaustion level in performance should be reduced or the performance should halt, else there would be a degradation of performance. On the other hand, the audience should be comfortable and at peace to enjoy the performance. The audience being at the receiving end should be comfortable as well and should not get bored with the performance; else they would leave the performance. Ergonomics as a subject should try to bring the artist and the audience close to one another by looking at the two ends of the system. This is a one-way communication and ergonomics should try to ensure a seamless one way communication by building adequate feedback. This feedback could be evident from the audience's body movement (they start dancing), clapping, or when they start repeating the dialogue of the artist. At times performances need to factor in the direct feedback received from the audience in terms of appreciation, facial expressions, emotional outbursts, etc.

1.8 THE NEED FOR MEASUREMENTS

A famous opera is coming to the town of "Bhimroti". People are flocking to the opera house in the town just for a ticket to get a glimpse of the fantastic performance of the opera. The name of the performance is "Telebhaja" which is a ballet and based upon a theme in the local context of the town where it's being performed. Then came the grand day of the performance. This was a newly built opera house and people were excited to see it as well as the performance. The problem started when people started entering the opera house just before the performance. At the entrance there was a complete chaos as not more than two people walking side by side could enter. Many people even bumped their heads while entering through the main gate into the lobby of the auditorium. The lobby was packed with people and people felt that the space inside the lobby was too cramped for the audience to stand. Some even started complaining that the authorities should have thought of the number of audiences.

People wanted to use the washroom and there also they faced problems. Many complained that the wash basin was placed at a very high level and it was difficult for them to reach for the faucet. It was creating problem for them in washing their hands. Not only was this, the mirror was also at a height such that one could see only a part of his/her face. After some time the audience were allowed to enter the auditorium and take their respective seats. People started complaining that the seats were too small and they were falling from the seat. Some even complained that their knees were touching the front seat when they reclined their seats. There were some who complained that they could not see the stage in front because of the people sitting in front of them. These were all problems at the audience end. The people were never thought of.

Similar problems were also noticed at the performers' end. The green room size was inadequate to accommodate all the performers and they were all grumbling that they were getting delayed in dressing up. The musicians seated in front of the stage and facing the audience started complaining that they were touching each other and it would be difficult for them to play their instruments. When the performance started even the stage area looked inadequate as it was difficult to accommodate all the performers on the stage and some even had to get down from the stage and stand on the ground at the same level as that of the audience. The performers complained that they could not see the people seated in the last row of the balcony. Even the dresses which were provided to them were so big that they were dragging on the ground making the movement of the performers extremely difficult. The performance ended, but it was a complete mess and the performers could not put in their best. The audience were extremely annoyed not by the performance of the group but because of the opera house and the amount of discomfort it triggered in all. Here also the people's perspective was never thought of.

The above example shows how human body dimensions play an important role in the designing of the opera house and its different components. Not only this, if the design of the work wear and the props is not in tandem with the users' body dimensions, then there are chances of misfits between performers'

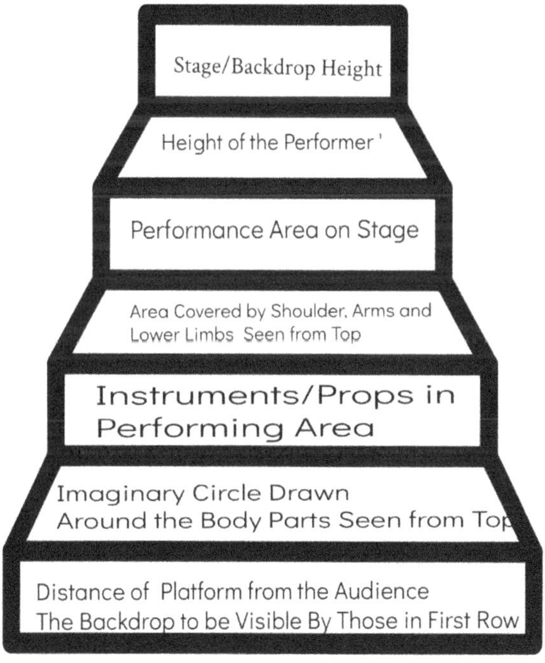

FIGURE 1.5 Human body dimensions in any performance for a seamless performance: A schematic representation.

props/ dress and venue and this affects the performance as well. On the part of the audience, a mismatch in the body dimensions and the space, seating, doorways and all those places where the human body comes in contact need to be designed considering different aspects of body dimensions and the movement pattern. Any performance that goes on at a venue should factor in human body dimensions and movements. If this is not done, the performance could just be a chaos as we have seen in the incidence above. In the domain of performing arts the human body dimensions (anthropometry) play an important role in organising the venue and ensuring that the required number of performers fit well in the space. It also ensures that the audience is also positioned comfortably and can see the venue clearly from all possible angles. Figure 1.5 presents schematically how body dimensions influence performance and ensure a seamless integration between the performers and the audience and hence ensuring a successful performance.

1.9 DIMENSIONS FOR DIFFERENT ELEMENTS IN PERFORMANCE

There are different types of performing arts and each has a different requirement or criterion for dimensions with reference to the human body dimensions. In general, for any performing arts there are certain common elements for which human body dimensions are essential. Some of these are mentioned below:

i. Handheld props: Handheld props which are used by the performers should be designed in tandem with the hand and body dimensions. For example, if a police baton (Figure 1.6) is designed, then the diameter of the baton should be designed factoring in the grip inner diameter of the human palm. You can get an idea about this if you hold the baton with your palm you need to curl the four fingers around the same with the thumb overlapping the four fingers. The diameter of the fingers is your grip inner diameter and needs to be equal to the baton diameter. The baton diameter should be a little less than the grip inner diameter so as to enable you to hold it with finger and thumb overlap. For the length of the baton you need to consider the palm breadth and then extend it beyond that to get to a decent length that looks good and easy to hold. To understand the palm breadth, spread your palm on the table facing upwards. The distance between the widest parts of the palm is the palm breadth. Sometimes the thumb is to be included.

FIGURE 1.6 Dimensions to be considered for a police baton. Photo by Kindel Media: www.pexels.com/photo/man-in-black-police-uniform-holding-a-baton-7714724/

If your prop demands that you need to hold it with the palm and direct it, then the thumb should be included. Example of this is holding and guiding an electric iron in a performance (theatre, movie, television set, etc.) (Figure 1.7). When you hold a baton you do not need the thumb to rest on the baton. Thus, in this case while taking the palm breadth the thumb should be excluded. For different handheld props, depending on the touch points of the hand, the respective dimensions of the body (anthropometric dimensions) need to be referred to. For example, if it's a walking stick (Figure 1.8), then to decide the height of the stick the hip height while standing could be referred to. For the handle diameter and length the grip inner diameter (referred to before) and the palm breadth (without thumb) can be referred to. So you need to be vigilant as to the type of handheld prop you are using. Any handheld prop which requires very tight grip and better control has to be held with the fingers curled and the thumb sitting over the fingers. The height of such props should be around the elbow height or a little below while standing or sitting. If the prop requires fine manipulation, then the fingers should come close together the way we hold a brush while painting.

FIGURE 1.7 Dimensions to be considered for an electric iron. Photo by cottonbro studio: www.pexels.com/photo/man-in-white-crew-neck-shirt-ironing-brown-shirt-5901631/

ii. Set props: Different types of set props used in performances should consider human dimensions as well. Beds should factor in the height of the person as well as her/his body width for deciding upon the length and width of the same. The height of bed, chairs or sofa set should factor in the dimension of the height of the region just underneath the knee to the floor. It should be equal to or a little less than this. This is because when a person gets up from a seated position of a chair, sofa or a bed, he/she needs to push the feet a little inside. Thus, when the height of the seat is low, this is facilitated. Handles of doors and cupboards should be at the elbow height of the performer to facilitate easy opening, keeping in mind to keep enough clearance between the handle and the cupboard. For calculating this the fist circumference of the user should be referred to and some allowance should be added to it for finger rings, gloves, etc.

iii. Dressing props: Dresses form an integral part of any performance and should ensure proper fit. For this anthropometric dimensions of different parts of the body are needed, with some allowance to facilitate movement of the body parts. For headgears we need the head circumference. For the headgear height we need to refer to height of

FIGURE 1.8 Dimensions to be considered for a walking stick. Photo by Alexander Zvir: www.pexels.com/photo/man-with-a-walking-stick-standing-on-a-paving-street-15757218/

the head curvature (Figure 1.9). If a helmet or a gear like mask has to be designed, then the dimension of the nose, ear to ear (Figure 1.10) needs to be considered. While designing headgears always remember that a good amount of body heats needs to be dissipated from the head. Thus, keep provision for heat escape in the form of slits or holes. In case of crown take the head circumference and fix it with compression clip on the two sides of the temple.

FIGURE 1.9 Dimensions required for designing headgears. Photo by Yogendra Singh: www.pexels.com/photo/people-celebrates-holi-festival-3367460/

For dresses for the upper and lower part of the body, two or more body dimensions need to be taken. For shirts and jacket-like wear, the bi-deltoid (Figure 1.11), forearm forward reach is to be considered. The length of the shirt should be decided by taking into account the belly so that the trouser belt is not exposed (Figure 1.12). Apart from this the chest circumference and abdominal circumference should be referred to. For trouser, the length from the hip to the feet, with allowance for folding the legs, is to be considered. The buttock circumference, the abdominal circumference and the crotch height are to be factored in (Figure 1.13). The circumference of the trouser should be in accordance with the feet length. In performances like magic,

FIGURE 1.10 Specific dimensions for designing helmets. Photo by Ron Lach: www.pexels.com/photo/mother-putting-a-helmet-on-her-kid-9964463/

where the dress needs to be changed very fast, extra allowance should be given for wearing the dresses one on top of the other, with the convenience of opening the dresses very fast after every performance in rapid sequence. In these cases the dresses should be a little loose fitting so that opening them rapidly becomes easy.

iv. Musical instruments: though musical instruments are of different types ranging from string instruments, percussion, wind, etc., changing their dimensions at random based on body dimensions might not be wise. This is because it might impair the sound emanating from the specific instrument. However, the point of contact with the users could be intervened. For example, heavy instruments like saxophone,

FIGURE 1.11 Dimensions required for designing shirts and jackets. Photo by Laryssa Suaid: www.pexels.com/photo/man-wearing-sunglasses-1667077/

which needs to be played with the fingers, need to be synchronised with fingers of small and large size so that they can have proper contact with the fingertip. The fingertip dimensions can be referred to for designing such additional "cap" like attachment over the existing saxophone keys. Other instruments similar to saxophone, which are heavy and lead to muscle ache and fatigue, need to be supported from the body parts with a broad fabric. This mechanism reduces muscle stress and delays the onset of fatigue. That is why Spanish guitars, saxophone, etc., are not held by hand and are suspended from the shoulder or neck with a thick fabric. The violin, on the other hand, is relatively lighter but is again supported at one end by the neck. The purpose of this is to ensure that it's steady and does not lead

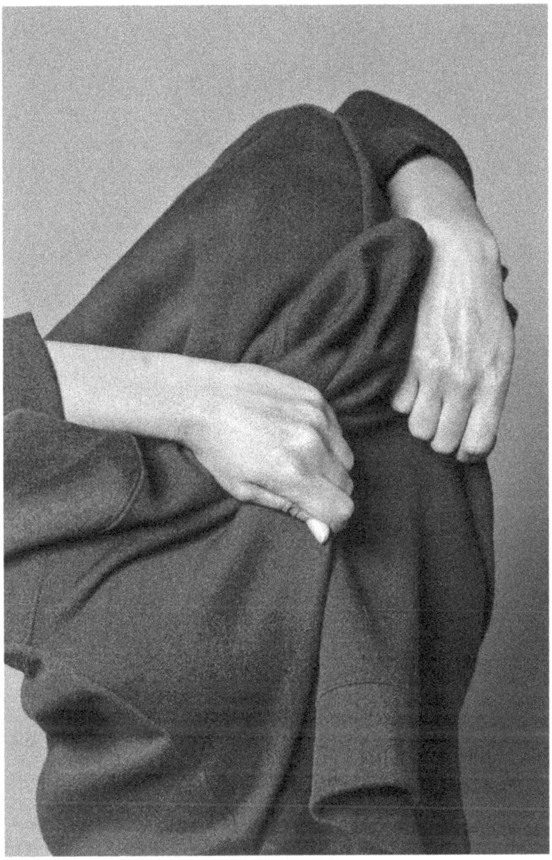

FIGURE 1.12 Specific dimensions of the shirt considering dynamic movements. Photo by Angela Roma: www.pexels.com/photo/ethnic-person-putting-on-dark-clothes-on-brown-background-7480096/

to load on the wrist of the non-dominant hand while playing it. For bigger instruments like base drum, the seating position of the player is important. As the player uses two sticks for striking on the drums and the tones, the height of the same should be at or near the drummer's elbow height while seating. For those playing the keyboard either while standing or while seating, since this is a light job, the keyboard height should be at the elbow height of the player. Thus, by making minute changes in the musical instruments the load on musicians is minimised and this ensures that the musicians are in a comfortable body position. This is one of the key elements for ensuring a good performance in any performing arts.

FIGURE 1.13 Dimensions required for designing trousers. Photo by destiawan nur agustra: www.pexels.com/photo/anonymous-man-in-black-boots-sitting-on-street-4315213/

1.10 OPTIMISING SPACE FOR PERFORMANCE

Any performance demands movement of the body. If a performer is singing some classical song, he/she might be stationed at a place, but there would be movement of body parts. In an orchestra where several musicians stand or sit close to each other there are movements related to playing the particular musical instrument. Similarly, in performances that demand movement like

FIGURE 1.14 Mapping the personal space of a performer. Photo by Valeria Costa: www.pexels.com/photo/woman-in-black-and-white-dress-reading-a-book-13607537/

theatres, rock concerts, magic, circus, etc., how does one get an idea about the space requirement?

You need to do this in steps. The first important step is to learn to "visualise" the space from top (called plan view) and the sides and rear (called elevation). The view from top helps you to estimate the space required by a person, or to estimate the imaginary space required by a person which is very personal to him/her and where no intrusion is warranted. This can be roughly drawn by taking the centre of the person's head and with this point drawing a circle of radius equal to the person's upper arm length (Figure 1.14). The upper arm should be stretched out at the sides, with the elbows folded. This circle is roughly the personal space of the person where no one should intrude; else his/her performance would suffer. While arranging for any performance one has to calculate that at a given point in time, a performer's personal space should not be compromised. This becomes all the more important if you have to calculate the space requirement for a number of performers. For example, if space has to be calculated for a large orchestra, then you have to calculate how many musicians would be there and then draw that many circles as mentioned above, side by side and with a little overlap to come up with the space requirement. If the musicians are using specific musical instruments, then visualise the performer from top to check whether the musical instrument is jutting out from

FIGURE 1.15 Musicians in in their personal space. Photo by Alena Darmel: www.pexels.com/photo/three-people-with-masks-playing-drums-7715851/

the personal space or not (Figure 1.15). If it does, then extra space should be allotted to the musician. This could happen in case of musicians playing piano and organs which are instruments occupying larger space and which normally extend beyond the primary space of the performer. If a musician is standing, the envelope diameter is smaller compared to when he/she is seating. Thus, for seated performers more space is to be added as in seating the personal space increases as depicted in Figure 1.15.

When it comes to performances on stage or on surface and involves movement, like in ballet, the space should also consider this dynamicity. The calculation of the space now becomes a little complex. The steps followed for

FIGURE 1.16 Space for orchestra. Photo by Ramazan Karaoglanoglu: www.pexels.com/photo/orchestra-and-conductor-during-the-performance-13673492/

the space calculation in case of orchestra remain the same, but extra allowance for movement needs to be added to the already static data to account for the dynamicity (Figure 1.16). In case of performances like circus, where movement takes place on horizontal and vertical surfaces, other considerations should be made. For example, the ladder used for climbing in trapeze act should ensure that the distance between the ladder steps should not be more than the knee height to facilitate easy climbing on a dangling ladder. As the ladder steps are used for holding while climbing, their diameter should be such that the fingers curl around the steps and the thumb overlaps ensuring proper grip for climbing. In the trapeze act when the performers stand on a platform high up in the tent, there also similar calculations need to be done for accommodating a requisite number of performers so that they do not fall off. The surface area of the safety net placed below in trapeze should factor the movement dynamics when the body falls. This is because when a person falls/jumps from the top the body does not always land straight but follows a trajectory. Thus, the area of the net has to be large enough to cover for the trajectory for every performer so that they do not fall outside the net (Figure 1.17).

In theatrical performances on stage, the distance between the wings is to factor in the elbow to elbow dimensions in relaxed state, as well as the shoulder width. The purpose is to ensure seamless entry and exit of performers, especially at a very rapid pace in tandem with the script of the performance. For movie sets also similar approach is needed in fixing the dimensions of the space and the different elements (furniture and other decorative and utility elements)

FIGURE 1.17 Area of the safety nets in circuses. Photo by Bonbon Lang-es: www.pexels.com/photo/white-net-bridge-across-forest-under-clear-sky-110315/

in the space. While designing any such spaces you need to trace the path of possible movements of the performers to ensure that there is enough room for all and while moving the body of one performer does not cover that of the other. This is an issue when an actor of taller stature is delivering dialogue in the front part of the stage and another actor of short stature is not active and gets covered behind the tall actor. If there is enough space, then the actor of short stature can move and reposition so as to make himself/ herself visible to the audience.

1.11 OPTIMISING ELEMENTS, PROPS AND ATTIRES

The ergonomic directions for optimising elements in space, props and attires have been discussed before. One should remember that in any performance it's a unique blend of the performers, space, elements in space, props and different attires. The dimensions of all these are extremely important to ensure the safety and quality of the performance. While these optimisations for dimensions are carried out, one has to factor unforeseen circumstances as well. For example, fires at the venue of the performances are often heard of. At times someone in the audience falls sick and demands immediate medical attention. In all these cases the exit doors should have adequate clearance for bringing in stretcher inside the venue if the venue is a closed one. In these cases there should be a provision for movement of stretchers. For fire and other incidences the crowd needs to be dispersed uniformly from the four corners of the venue to avoid any delay in evacuation and reduce the incidences of a stampede.

1.12 THE STEPS IN OPTIMISATION

There are certain steps in the optimisation of dimensions for space and props. They follow a common pattern to a certain extent. The steps for optimisation are roughly as follows. You need to keep in mind that the data base for anthropometric dimensions (human body dimensions) is available with us. If this is not available, then we would have to measure dimensions of different types of target users, that is, users who are relevant in the context of the performance comprising both the performers and the audience (Figure 1.18).

Step 1: Identify the touch point for props, space which is in touch with the user (performer and/or audience).

FIGURE 1.18 Measuring performer's body dimensions. Photo by Kosygin Leishangthem: www.pexels.com/photo/group-of-men-holding-drums-performing-on-stage-2888802/

Step 2: Try to find out which part of the body dimension is relevant for that touch point.

Step 3: Write down those body parts.

Step 4: If you have access to data base for body dimensions, refer to the different values presented as percentile.

Step 5: Two principles normally apply. If you have to look for clearance, then take values which are on higher side. For example, if you have to fix the height of the door, fix it at a higher level for the taller person to pass through and the shorter person would not have any problem. If you have to look for reach, for example, keyboard height, fix it at the elbow height for the shorter person, the taller person can still manage by stooping a little.

Step 6: If you have no data base for body dimensions with you, you have to measure the target users and create a data base for different parts of the body relevant for the space or props. Then you have to apply the principles as mentioned in step 5 above.

The same steps also follow in the optimisation of spaces (Figure 1.19). You just need to visualise the space from top and identify the touch point of the

FIGURE 1.19 Optimising a space for performance. Photo by Harrison Haines: www.pexels.com/photo/photo-of-man-playing-guitar-3536236/

users with the space and props. Your focus should be on the head, shoulder and arms of the users and their movement in space and interaction with different props. For example, if you visualise a performer playing the Spanish guitar from the top, you would notice the head, part of the shoulder and forearm and a large part of the guitar protruding out from the body. This person now needs a larger space because the jutted portion of the guitar needs to be accommodated in his/her personal space as well.

1.13 ACCOMMODATING THE EXTREMES AND MOVEMENTS DURING PERFORMANCE

It is not necessary that your audience and performers would be of "so-called" average height and weight. They might be very tall and obese as well. You cannot deny entry to an obese audience. Similarly you cannot say "no" to a very tall performer for any performance for any musical extravaganza or theatre. At times the performance might also demand people with extreme dimension. So how do we accommodate them? Important touch points should be designed for the extremes. For example, if entrance and doors are designed for extremes, there is no harm. Handheld props could be made adjustable so that extremes can use them as well. In case of prop wears extremes or extra-large sizes should be kept handy, because adjustability here would not ensure good fit and might not be always feasible. Slip-on could be made of adjustable length by providing elastic band between the heel and the forefoot. But this might not be feasible for a performance which requires vigorous movements. Here the shoe might come out. Similarly crowns used on heads of different dimensions could be reinforced with elastic band for accommodating heads of different circumference. In case of hats this approach might not work.

1.14 AUDIENCE VISIBILITY IN PERFORMANCE

You often hear your friend complaining that he could not enjoy a performance because there was a very tall person sitting in front and he had to look at the stage from the sides (Figure 1.20). This is where the auditorium design and the design of seating system have to be looked into. Each person has a visual

FIGURE 1.20 Visual cone in audience seating. Photo by Tima Miroshnichenko: www.pexels.com/photo/people-sitting-on-red-theater-seats-with-3d-glasses-7991581/

cone. This is calculated by taking his/her height of the eye from the ground. From the eye a line is drawn parallel to the ground. An angle of roughly thirty degree above and forty-five degree below is drawn to give the vertical visual cone. Similarly there is a horizontal visual cone drawn from top view. An angle of fifty-five degree on either side of the perpendicular (in top view) makes it around one hundred and ten degrees. While designing the seating arrangement you have to calculate the inclination of the surface (might call it a gallery if it's indoors, but in outdoor arrangement you have to use mud to wooden plan to create this). The inclination has to be fixed by considering the tallest person sitting in front and the shortest person sitting behind and keeping on increasing the tilt of the surface on which chairs are placed till the shorter person's cone of vision is clear and unobstructed. For this both the vertical and horizontal cones of vision need to be checked.

1.15 KEY POINTS

I. The macro picture comprising smaller components making up the bigger picture is important in performing arts.

II. The performer and the audience are two sets of users and there has to be a seamless communication between them.
III. Improving performance is dependent on how well we factor ergonomic issues for performers and audience in terms of their characteristics, capacity and limitations.
IV. Performance is also affected by surroundings which include people as well.
V. Good performance is dependent on the context, artists' well-being, audience comfort and good communication with the audience.
VI. Visualising the context of performing arts as a larger picture with smaller components, which comprises the audience, performers, props, stage, lighting and acoustics, helps in identifying the ergonomic issues.
VII. Human body dimensions need to be considered in designing performance venue and props.
VIII. Take maximum value when considering clearance and minimal value when considering access.
IX. Always consider movement of the users and add extra allowance when needed.
X. While deciding upon the total space of the performance areas consider the number of performers and their props/instruments as well.
XI. For extreme body dimensions in props you can think of adjustability.
XII. To ensure everyone is able to see the performance calculate the inclination of the seating surface with the shortest person behind the tallest one and calculate the visual cones.

1.16 PRACTICE SESSION

i. A musical performance has been organised at a venue which is an open paddy field. A famous singer Mr Pagla is going to perform along with his team of musicians. You have to outline the ergonomic issues that need to be focussed on for ensuring the performance is a grand success.
Ergonomic Directions

A. First visualise the bigger picture and then try to identify the smaller elements that make up the bigger picture.
B. Try to identify the vulnerable points in this bigger picture where things could go wrong.
C. Apply ergonomics to make the vulnerable points robust.
D. Think of how two-way communications could be established.

ii. A circus company is coming to your town. They have identified a field and are looking for your inputs in terms of ergonomic for enhanced performance and better business. What would be your advice to them?
Ergonomic Directions

A. Selection of the venue with easy access from all around
B. The macro picture mapping of the space when the event starts
C. Seating position of the audience around the ring
D. Equal visibility and sound for all audiences

iii. A street play is being organised in one corner of a busy market at night. You have been asked to optimise the space to ensure best performance by the group. What would be your approach for the same?
Ergonomic Directions

A. First, list down the maximum and the minimum number of performers at any given point in time at the venue.
B. Based on this number calculate the individual spaces.
C. Add to this extra allowances for movements and props.
D. Demarcate audience zone keeping at least a distance of four hands from the performance.

iv. A ballet on ice is being organised at a makeshift venue. Your advice as an ergonomist has been solicited for both the stage and audience sitting. How would you approach this?
Ergonomic Directions

A. Get the maximum number of performers.
B. Add extra allowance as on ice there would be vigorous movement.
C. Add extra space for props if any.
D. For audience seating consider the shortest person sitting behind the tallest one and change the seating inclination accordingly.
E. Make sure that there are exit points at four corners of the venue to ensure uniform dispersion of the crowd.

Posture in Performance and the Role of Props and Instruments

2

2.1 OVERVIEW

This chapter gives an overview of the importance of body posture in different aspects of performing arts. The readers are taken through different types of body posture and how to identify the right and wrong postures. The relationship between different postures and how to ensure proper postures while using props during performance are discussed. The issues of fatigue related to posture in performance and the importance of adequate work–rest cycle in performance are addressed in this chapter. This chapter also gives the readers an overview of the application of human factors in designing and the use of different props and instruments in any performance. The chapter takes the readers through the human factors issues of ensuring proper fit of props and instrument to the relevant part of the human body. The issues of look and feel of props and instruments, their specific weight and how each affects performance have been discussed. The human factors of props with reference to the performer and how these affect the mood of the show are covered in detail. The role played by the weight of props and different musical instruments and how it can influence performance are discussed. The role played by props in the form of attire while moving during performance and the human factors angle of rapid changing of body props or attires are also discussed in detail in the chapter.

DOI:10.1201/9781003506409-2

2.2 WHY BODY POSITION IS IMPORTANT?

Bitto is a professional vocalist. He is performing at a college fest for the past three hours, with small breaks in between. He stands and sings at one place with the microphone fixed to the stand as he is more comfortable with this position. The audience is enjoying his performance, they are singing with him, clapping, crying in emotional songs and at times are also dancing with his song. Things were going on well. Suddenly Bitto stops singing and falls on the stage. The audience is worried as to what happened to him. He is taken to the nearest hospital, where the doctor examines him and declares that it is nothing serious. Bitto gains consciousness. The doctor says that since he was singing while standing at one place, there was movement of blood from the upper part of the body to the legs. This led to shortage of blood in the brain, which ultimately led to Bitto's collapsing. This incidence points to the fact that in performing arts artists need to take care of their body position while they are performing, else their performance might not be good, or they might fall ill or suffer from discomfort in their body parts.

The body maintains certain positions while performing any activity and also while standing or sitting still. Body position plays an important role in ensuring that performers give in their best and continue performing for a longer period of time. A good body position often called as a "neutral" body position ensures that the body parts, especially at the joints, are at a specific advantageous position where fatigue would be delayed. This could be at an angle of ninety degree or more at the elbow or at the knee. This could be maintaining an erect spinal position. These ensure that the body parts function optimally at these angles and they can exert maximum force and the controls with their respective body parts are also very precise. Not only do these body postures delay the onset of fatigue, especially in any high-activity performance for a longer duration, but they also ensure more spontaneous movement of body parts while performing. Saying that maintaining an erect back for a long duration at a place is not possible and the performer might get fatigued and this might affect his/her performance.

Therefore, the performers apart from adhering to the above norms for different aspects of posture should ensure that there is adequate change in posture. This would ensure that the fatigue is delayed and the performance is increased. One needs to remember that neither remaining static in one position nor vigorous movements in awkward position of the body like bending forward or sideways, twisting the upper part of the body, reducing the angles between the joints or working or holding heavy objects away from the body should be avoided at all cost. Every effort should be made to change posture at frequent interval and if working in constrained posture is mandatory, then micro breaks should be taken. These would ensure optimal performance.

A theatre was being staged in one of the auditoriums in the town of "Bhondul". The show was progressing very well which was evident by the loud round of applause from the audience. There was a scene of an elderly person who is in a retirement home. He sits for the whole day inside the room and reminisces all that he had done through his life. He is aged, cannot sit erect and cannot move. He sits on his bed and thinks of the past. This particular scene was enacted by a young actor and it lasted for an hour. The expression on his face and his dialogue delivery were just amazing. Suddenly the audience finds that the actor is in intense pain and he falls from the bed to the stage and starts yelling. The audience starts clapping thinking that it's a part of the acting. The show stops and the paramedics rush to the spot. The actor is rushed to the hospital emergency. Investigation revealed that he was suffering from intense back pain. This was mainly due to the fact that he was sitting in a forward bending position and for one hour. Moreover, this was his tenth consecutive performance and he didn't get any rest. This accumulated and finally flared in the form of a debilitating low back pain. Thus, awkward postures if mandatory in a performance should be interspersed with micro breaks. If possible some movement should be included. This performance of three hours duration did not have any intermission which probably added to the incident and escalated it further.

2.3 DIFFERENT TYPES OF POSITIONS YOU MAINTAIN WHILE PERFORMING

Ms Bikemonto is performing on stage in a seated position. She is a vocalist and is seated in a cross-legged position on a platform on the stage. She is accompanied by a percussionist, violinist only and she herself is singing while playing the harmonica. She is singing for the past four hours, but she and her team are taking small breaks wherein they are moving on the stage and again sitting down to sing in the same cross-legged position on the stage. This type of change in position of the body is the key to reduce fatigue and carry on the performance for long with equal energy all through the concert, lasting for three to four hours and at times more.

Performances are of different types, with some remaining static and some involving vigorous dynamic movement. A circus performance is in progress. Two young performers are walking over a tight rope. Suddenly another performer comes with his bicycle and starts riding over the tight rope. A close look at the performers reveals that their position is constantly changing in order to balance themselves on the rope. Even the performer riding the bicycle has to tilt from side to side to maintain balance. This is an example where there is constant change in posture but for a brief period of time. As the time duration is too short, these constant changes in posture are beneficial for the body and the performers do not suffer from early fatigue. If these activities have to continue for three hours or more at a stretch, the performers would have been tired and fatigued very easily.

A vocalist is performing on stage. She is seated on a high stool and singing while playing her Spanish guitar. She is accompanied by a bongo and a Congo player and by another flute player. The performance is to continue for four hours without any break, except for drinking water. A performance like this is always strenuous for the performers as they are confined to a place for a prolong period of time without any movement. Moreover, the vocalist is just seated on a stool without back rest and she has to maintain an erect position of the spinal cord, which is extremely difficult. After some time it really becomes difficult for the vocalist to perform as she starts suffering from back pain. With pain in her lower back she is unable to concentrate on the performance and the show suffers.

FIGURE 2.1 Ballet on ice with different dynamic positions of the body. Photo by Pixabay: www.pexels.com/photo/women-s-dancing-ballet-358010/

A ballet on ice is in progress. There is a perfect synchrony between the performers. They are performing on ice at a rapid speed with perfect synchrony. In such cases there are frequent position changes in the body and hence little changes of pain in body parts, except for a little fatigue after the performance. No ergonomic suggestions are warranted here. The only suggestion in terms of posture is to ensure that dynamic postures are maintained and there is no awkward posture involved. In performance one should be careful in maintaining a dynamic posture and avoiding awkward postures for longer than twenty minutes or so (Figure 2.1).

2.4 THE EXERTION WHILE PERFORMING AND HOW TO HANDLE IT

Mr Bakwas is a well-known theatre personality. He performs simultaneously in three different theatres, and the audience wonders how he manages to remember the dialogues. His performances on stage are full of actions and he carries them throughout the play for the entire duration of three hours. He is always in the lead role. Interestingly Bakwas is never tired as he takes adequate rest at night and his performance is interspersed with mini breaks in between. This helps him to recover and the onset of fatigue or tiredness is delayed to a great extent. Thus, he knows how to handle his exertion. In fact he has to stoop forward in one of his plays for almost the entire duration. So he ensures that when he goes home he not only switches off his mobile phone, stops talking but also performs some reverse bending exercise and thus he does not suffer from low back pain or other ailments. Bakwas knows how to handle exertion in his performance.

Mr Pagla is a vocalist. It's the peak season for performing artists like him. He has four concerts today and each of his concerts lasts for around three hours, because he has to respect the request of the audience. He has three accompaniments with him. He himself is seated in cross-legged position on the Dias and plays the harmonica as he sings. The others are a percussionist, a keyboard player and a piano accordion. The show was going on in the last venue of the day and everyone was putting in their best. The audience was appreciating and enjoying the melodious performance. Suddenly the percussionist lost tune and the entire synchrony of the musicians was lost. It was found that the percussionist was getting slower and could not keep up with the rapid pace of the performance. Pagla asks the percussionist Ajoy, "What's the problem Ajoy"? To this Ajoy replied "Sir my hands are aching and I am unable to play anymore. I need some break". A careful analysis of the situation revealed that Ajoy was involved in the task that demanded light force but high repetition involving the fingers and the palm. This was being continued for the past five to six hours in the day. These activities lead to very small wear and tear in the tissues of the hand. With adequate rest they recover. When there is no rest as was in this case, it piles up and flares up as a major problem in the form of pain and discomfort. The only remedy to such problems is to ensure adequate rest lasting for seven to eight hours to allow the body to recover and fit for work once again. Thus, performers as a rule should be instructed not to exert more than fifty per cent of their maximum force. In this case the force was the striking force with the fingers. To get an estimate of fifty per cent of the force first exert your maximum force and the next time you exert force reduce your force by a little. If possible the musicians could also change their positions. This ensures that every team member changes his level of exertion. Thus, the percussionist shifts to piano accordion, the piano accordion shifts to percussionist and so on. Though this might not be always possible, this is another good way of evenly distributing the load on all without putting the stress on a particular musician or performer.

2.5 THE IMPORTANCE OF REST IN BETWEEN PERFORMANCE

Ms Hasi is only twelve years old and is performing in the first round of a reality show. She is singing a very popular movie song and everyone is listening to her in rapt attention. She has a very melodious voice. After singing her second song she suddenly collapses on the stage. She is taken to the hospital and later on recovers. The doctors upon examining her found that she did not have good sleep for the past two days and was practising for this reality show almost twenty-four hours. She and her parents thought that by putting in such extra efforts she would be able to perform well. It doesn't happen in this way!!

In performance the body works and burns calories. If the performance is for an hour, then for experienced performers it's not a problem. Whenever the performance demands vigorous activity like acrobatics in circus, ballet, dance, etc., there should be a break of a few minutes after forty-five minutes to one hour of performance. This ensures that the body recovers and is fit to resume in the next bout of performance after the micro break. Thus, work–rest cycles are an important means of ensuring that performers are not overloaded and the performance is not affected. The key to any good performance is practice and also small micro breaks in between. The breaks could be small, of a few minutes duration, or could be larger breaks of thirty minutes duration like it happens in theatres, operas, etc. If performances last for more than three hours, then the break duration should be more and if needed two to three intervals may be introduced. The main objective of rest or pause is to give the body time to recover and start with increased vigour.

2.6 YOUR PROPS AND INSTRUMENTS IN POSTURE

A group of musicians are finding it hard to fix their equipment at the proper height for a musical performance. The drummer is unable to adjust his seat at the proper height, the tabla player doesn't know at what height should he place the tablas, and the artist who is going to sing while playing the harmonica is confused with regard to the appropriate height as well. The entire group is new to this profession and has just started its stage performances. Suddenly an elderly musician comes to their rescue. See friends…your instruments' height has to be fixed depending on certain reference points in your body. He says three reference points are to be referred to: the chest, the elbow and the waist. If you need to exert force, then place it at your waist height or below it. If the force exertion is very light, keep it at your elbow height while sitting or standing. If the instrument requires only feather touch, keep the height close to your chest height … Saying this he smiles and leaves the stage. The musicians are amazed at the person's prudence and capability of solving this problem.

Props both handheld and wearable are essential part of any performance. You should only ensure that the prop does not lead to any awkward posture, else it would lead to fatigue and degradation of performance. Handheld props should be placed at appropriate height to ensure erect posture. If light work is to be done, then props in space should be placed at the elbow height. If precision work has to be done, then the prop should be near the chest. If heavy work has to be performed, then the prop should be positioned near the hip where you wear your trouser. Any handheld prop should ensure power grip with the fingers curled around and the thumb overlapping. Care should be taken to keep your wrist more or less straight and not bent. The arms should be close to the midline of the body (imaginary line dividing the body into two halves which are mirror image of one another). The upper arm should be kept close to the body for the major part of the performance to ensure that pain, discomfort and fatigue in different body parts are delayed. Make sure that the performers are not carrying very heavy props and holding them in air without any support. This would also trigger early fatigue and lead to pain and numbness in body parts. Whenever heavy props are to be handled, they should be supported either on the ground or should be made extremely light so that they can be lifted and kept in air by a few fingers.

2.7 POSITIONING THE EQUIPMENT FOR PROPER POSTURE CONTROL

Akaj is a pop singer. He is singing with headphones and making vigorous movements as he is singing. He is almost jumping from one end of the stage to the other end. Suddenly while doing this his headphone and microphones dropped on the ground and he is completely embarrassed on the stage!!!

All the equipment associated with performance should be positioned in a manner that performers do not fall and are able to maintain their body balance. Headphones and microphones fitted on the ear and the head should be light and the performer should be able to move the head with ease. Props in the form of set design should be so arranged that performers are stable on the ground. Musical instrument should be firmly placed so that the performers have a good command over them without the need to lean, stoop or assume any awkward posture.

2.8 FITTING YOUR PROPS AND INSTRUMENTS TO YOUR BODY

Body props as has been discussed before should ensure good fit with the performers. An actor was once wearing a very loose fitting robe of a priest. While performing on the stage he tripped and fell and thus injured himself seriously. Not only this, the performance was also affected and the audience was extremely annoyed. Thus, props and especially attires should fit the body and need to be customised for individual performers. In case of performance requiring vigorous movements, there needs to be adequate allowance for your attire. If it's tight, movement would be restricted and if loose performers might fall down due to trips, or the loose end of clothing getting entangled in nooks and corners or into some protruding objects. In case of props which performers hold, you need to be careful that they fit the contour of the body and do not protrude out from the body. If it does, then controlling such props would be difficult. For musical instruments, you cannot change the size or the form. As such instruments require very fine dextrous movement of the fingers and the palms, they should always be located closer to the body for better command. If possible and without compromising on the acoustics of the instruments, additional attachments like a string, stand, etc., could be used. For example, if the performance demands that the performer has to move while playing the Spanish guitar, then the use of a string to hang it from the shoulder is better. If percussion instruments have to be played while standing, then they need to be placed on a stand to ensure that the musician is standing straight and not stooping forward (Figure 2.2). It needs to be ensured that the instrument is placed at the elbow height or a little below for better performance. You have to remember that your props and instruments are an extension of your body parts. These give you a separate identity. While acting the props help you to immerse yourself into the character that you are trying to enact. In case of

FIGURE 2.2 Heavy instrument (Irish Harp) being played while resting on the ground. Photo by Burak Esen: www.pexels.com/photo/woman-playing-the-harp-on-the-beach-18698306/

music whether you are playing the violin or the harp, they in essence represent your talent in producing that melodious sound appreciated by your audience. For heavy instruments or props it is advisable not to move during performance as it would unnecessarily increase performers' energy expenditure and expedite the early onset of fatigue.

2.9 THE LOOK AND FEEL OF YOUR PROPS AND INSTRUMENTS

In any performance the look and feel of the ambience plays an important role in the success and failure of the performance. Thus, props and instruments should ensure that while performers use them there is adequate "feedback" from them as it gives them an assurance that they are in proper position and thus are in control in case of handheld props or in case of musical instruments which are solely operated by the hand. It's also important that the very look of the props and instruments should be in tandem with the ambience. Just think of a musician playing a bright yellow violin. Your attention would go on the weird colour of the violin rather than his performance. If it's a very romantic mood, then this yellow colour would come in the way of that mood in the performance. Imagine someone is playing a piano sitting on a pile of bricks on the stage!!! How would you feel? If the mood of the performance is "romantic", then all attention would be on the pile of bricks rather than on the music itself!!

Mr Haridas is playing the role of a father in a family. He is rich and the Chief Executive Officer of a multinational company. He is shown entering his office in a red coloured shorts and yellow shirt, with white sneakers!! The very attire of the actor disrupts the mood of the play and the audience instead of taking it as a serious scene start laughing. So the look and feel of the attire should match the mood of the scene which is being enacted. This helps in creating a mental image among the audience. Thus, props both handheld and wearables need to be designed in tandem with the mood of the performance. If this is not done, it might affect the performance. Other props on stage are also very important. When the audience see a scene, they scan the entire stage in its totality. This includes all the elements on the stage like the tables, chairs, etc. Thus, it's important that all the elements on the stage are arranged and designed to set the mood of the performance. They should not be an odd man out. For example, in a vintage scene the furniture should also be antique and not modern!!

2.10 THE WEIGHT OF YOUR PROPS

In a movie shooting an actor was playing the role of a champion weightlifter. The director advised him to lift a prop made of wooden frame which mimicked the weight. The actor was too ambitious and thought that it would be more realistic to lift the actual weights. At the very first take when he tried to lift the weight he fell down and injured himself. In a similar incident, the maiden show of a play was enacted. The scene involved a fighting scene between two people with sword. Both the players wanted it to be realistic and ordered for the real sword in the maiden show. When the scene came one of them could not lift the sword because it was too heavy for him. The show turned out to be a joke as the fighting scene could not take place because one of the actors was unable to lift the sword leave alone the fighting scene. Carrying a very heavy handheld prop throughout the performance leads to early onset of fatigue and can degrade performance. Apart from this heavy props when carried can lead to pain and discomfort in different body parts and this can linger for long even after the weight is released. Thus, performers need to be careful in not carrying any heavy loads. If they have to, they need to exercise daily with the equivalent amount of load to ensure that the muscles are toned accordingly and there is no pain or discomfort in different body parts. This requires time to adopt and that could range from a few weeks to a few months. Such heavy props also restrict movements while performing in any medium.

2.11 PROPS WITH REFERENCE TO THE POINT OF REFERENCE IN THE BODY

The props used in any performance are meant for enhancing or augmenting the role of the performers or the performance in general. In a set either for movie shooting or in a theatre the props give the audience a flavour of the context of the scene. For example, a bedroom should contain a bed, side table and appropriate lighting. A dining room should contain the dining table with plates, knives and forks. All the finer details of the props are extremely important in conveying the theme of the particular context to the audience without the need to narrate it verbally. Performers (or actors) are a part of this context who make the context lively through their acting and dialogue delivery. Thus, this is a systems perspective, where there is a bigger picture which is the set and smaller elements that make up the bigger picture which are the props and the actors as well. The props thus need to be properly positioned in the space with specific reference to the pathway for movement of the actors so that there is a perfect integration of the props and the actors and they look like a unified whole making up the entire scene which is being enacted. No props should look disproportionate and out of context. If it does, then the stage which is a platform for the scenery being enacted would look distorted to the audience and they would not be able to connect what is being conveyed to them.

Audience develop a model or imaginary picture when they watch a performance. This picture gradually gets established in their brain and as they get deeply engrossed in the performance, their expectation increases and they expect more. This is how the climax of any performance is reached and if the performers are able to hold onto this tempo, the performance becomes successful. The props are an integral part of this performance and they help in augmenting the performance by bringing it close to the reality.

2.12 ATTIRE AND PROPS WHILE MOVING

When performers move on the stage they do so along with their props and attires. At times the props on the stage which add value to the scene or the context have to remain stationery as they do not move due to their weight. It is to be taken care that when performers move around heavy stage props, there should be no projections or sharp edges, else they can get caught or the attires and handheld props can get caught and the performer's normal flow would be restricted. The attire and handheld props should be light enough to ensure that the performer's movements are not restricted.

Ms Dikhchao is a dancer and is performing today on an open stage in a rural setting. The lights and the acoustics have been arranged accordingly and the sitting arrangements for the audience have been made all around the stage. She is performing with her entire troupe of ten more performers. There are some props on the stage to give the outlook of an ancient place of worship. These comprise a banyan tree, a cut-out of a cow, two huts and a bullock cart. The performers were carrying fire torch in their hands with live fire in them. They were all wearing a special clothes called saree which actually is wrapped around the body. The performance started well. The audience were mesmerised by the performance and the speed of the performance. There was a scene of four performers jumping on the bullock cart and dancing. The moment the performers jumped on the bullock cart, the cart moved as it was not fixed to the stage. This destabilised the performers and they fell on the stage with the burning torch light in hand and immediately the stage was on fire. The audience started screaming and running away and there was panic all around. The fire was put off but the show had to be stopped because the fire had already damaged a large part of the venue and even the performers had sustained burnt injuries.

The above incident is an example that when movement takes place on stage one has to factor in the weight and the nature of the handheld props and the props on the stage. The props on the stage, in this case bullock cart, were not designed or fixed in accordance with human movements. All the performers jumped on the bullock cart from one direction, leading to its movement. They were holding hand props which had live and naked flame. The cascade of events with props in hand led to the disaster. This is where one needs to think of the bigger picture and the smaller components that make up the bigger picture in any performance. Human factors approach helps you to visualise a performance in these situations and helps you to predict where and how things can go wrong. This paves the way for remedial measures as well. Thus, one needs to keep in mind that things can always go wrong and it's wise not to believe that everything will be fine…

2.13 RAPID CHANGE OF ATTIRES AND THE HUMAN FACTORS ISSUES

Ms "Arnei Dorkar" is performing in a one act play where she is the lone actor and she depicts the role of a woman from childhood to death. At every scene her dress changes and the change is very fast, almost in the blink of a second. The audience is mesmerised by her performance and they are wondering how such rapid change of dress is possible. Her show is a super hit. Human factors have a role to play here as well.

In magic shows and other forms of performances rapid change of dress is needed. The change is so rapid that it happens within minutes. The lights dim or the performer exits through the wings and comes to the stage in a new dress all together. The key to performing this rapid dress change is manifold. One of them is wearing of clothes in layers so that one might simply strip off one layer and the new layer is exposed. For doing this the layered clothings need to be attached with Velcro strip to facilitate easy opening. A common point of opening is tricky because once pulled from that end while in between the scenes the performer is already under stress and might pull out the subsequent layers as well. So it's better to have the opening point at different locations but at the same reference point in the body. For example, if the hip is located as the point from where the clothings would be stripped off in layers, it has its advantages. The biggest one is its proximity to the limbs (hands). So the point of opening could be around the waist at different points till the hands are able to access them. The opening points should be demarcated by texture and bright colours so that there is a double reconfirmation to the performer about its exact location from where it has to be pulled without looking. If needed small numbering could also be done to aid in this process. In cases where different types of dresses have to be layered, for example, jeans trouser and shirt and skirt, in that case the jeans trouser should be layered first and the skirt next. For easy opening of trousers the opening of the trouser could be made with Velcro strip near the hip joint. Extreme care should be taken to ensure that wardrobe malfunction does not happen in performance.

To further facilitate rapid dress changes, the design of the fabric is important. Motifs should be used to camouflage joints, layers, and opening points on the garments from a distance. These create optical illusion and make the attire looks like a single piece. The optical illusion from the viewpoint of the audience does not allow them to distinguish the layered clothing, joints, etc., as the eye scans the attire as a unified whole. You must have seen magicians performing on stage, changing their attires in every scene. This is possible if the above human factors approach is followed. Only keep in mind that the opening should be within easy reach of the user, and the point of opening should involve using all the fingers. Care should be taken to ensure that pinch grip does not happen as that strains the fingers and forearm muscles. The use of all the fingers should be facilitated. Special care should be taken while washing these clothes. All the layered clothings should be of the same materials so that while washing they should not shrink, crumble, etc., and that would not allow it to settle one on top of the other. In case it does, then the layered clothings should be ironed to ensure that they fit over one another.

2.14 KEY POINTS

I. Stable body position is important in performing arts.
II. Performance demands change in body position and this should be maintained.
III. Avoid static body position.
IV. While using props ensure that the body is erect and the limbs are close to the body.
V. The wrist should be kept straight while holding handheld props.

VI. All props, instruments and equipment should ensure that movement is possible and the body remains stable with no awkward posture.
VII. Props weight should be easy to lift and handle.
VIII. The handheld props should match the attire, not necessarily gel with it.
IX. Stage props should ensure that performers' movement is not impaired.
X. Rapid change of attire could be achieved through layered clothing.
XI. The musical instruments while performing should fit to the body and if too heavy should rest on the ground, or hung from the body parts.

2.15 PRACTICE SESSION

i. For a dance drama to be performed what ergonomic suggestion would you suggest to the group and the performers for ensuring maximum efficiency?

Human Factors Directions
A. No performer to stand at a place for long.
B. The performance should ensure movement for all.
C. In case the performance duration is more than two hours, a brief intermission of fifteen minutes is to be given.

ii. An opera performance is taking place. It's a four-hour performance with around 90 performers on stage at different times, including the orchestra. Your ergonomic advice is solicited for the best performance.

Human Factors Directions
A. Every performer should have adequate rest the previous night.
B. There should be only one performance in the day.
C. If possible performers should be asked to do light exercise in the morning such as brisk walking and jogging to tone their muscles.
D. There should be at least two intervals of fifteen minutes in between.
E. No heavy handheld props should be used

iii. A ballet is organised in an auditorium. There are a number of handheld props to be used. The props are mainly big boxes to be carried by the performers and they need to dance with them. What human factors issues would be looked into to make this performance a success?

Human Factors Direction

a. The attire should be designed in such a way that carrying of the boxes would be easy.
b. Tight fitting attires would ensure better movement.
c. Boxes are to be very light with facility for gripping with all the fingers. If possible attach them with a string to the chest in a cross-wise manner like a Spanish guitar.
d. The area of the stage needs to be measured to ensure that the performers along with the handheld props are able to move with ease.

iv. A magician is performing. This is his debut performance. He has to change his clothes 20 times and that too within a very short span of time. Your human factors expertise is solicited.

Direction

a. Layered clothing could be thought of.
b. The opening of the clothes could be at the waist level.
c. Clothing and layers should be planned for ten scenes before interval and ten after interval.
d. Check for movement after putting on the layered clothing.
e. Use patterns on clothes and if possible small reflectors or glasses.

Illumination and Sound in Facilitating Performance in Different Mediums

3

3.1 OVERVIEW

This chapter gives the readers an overview of the human factors issues that help in creating an ambience wherein all the senses, mainly the ears and the eyes, are used. The chapter presents different aspects of a performing ambience and how to create it. The aspects of audience reaction to the performers' performance are discussed. The use of disappearing acts on stage and voice modulation for grabbing audience's attention are discussed here. The ways and means of voice modulation and ensuring that sound is used judiciously not to create noise are discussed. The readers are taken through different human factors issues in film and video as a communication medium. The chapter analyses the different issues in recording and playing the same. It discusses the role of human factors in building a storyline and designing the set in tandem with the same. The chapter tells the readers how to make judicious usage of colours. A major problem of distraction from the crowd while

DOI:10.1201/9781003506409-3

shooting any sequence and the human factors approach towards the same are also discussed.

3.2 MULTISENSORY AMBIENCE CREATION

In any performance the ambience is created by the usage of multiple senses. The multiple senses include the ear, the eyes and sometimes the nose. The feedback received by the audience through these senses helps in generating or creating an overall environment that helps the audience to enjoy the performance. For example, a performance which depicts the whole essence of a garden in the evening surrounded by blooming flowers possibly needs depiction through proper lighting, sound and if possible a perfume to create the ambience. Thus, three senses of the audience are being utilised, namely, the eye, the ear and the nose. This utilisation of more than one sense in humans for creating ambience is the main factor in creating a multisensory ambience. This is to ensure that the audience becomes a part of the performance which is being enacted. To make this more realistic the performers come close to the audience and then move to the stage. In operas the musicians are seated in front facing the audience and not on the stage, but at the same level or on a slightly raised platform. This facilitates direct eye contact with the audience and being at the same level does not create a sense of hierarchy and the audiences also feel connected with the performance. The biggest yardstick for a good performance is when the audience does not utter a word and remains engrossed with the performance for the entire duration.

Mr "Leto" the famous vocalist is performing on stage. The song relates to the advent of spring season. As he sings the illumination on the stage changes and on the screen there is a projection of different flowering plants. All of a sudden along with the song where "Leto" sings "the smell of rose…."… the entire auditorium is filled with the fragrance of rose. The performers have synchronised spraying of rose perfume with the lyrics of the song (Figure 3.1). The entire audience is mesmerised by this performance and they feel that as if they are sitting in the rose garden and listening to "Leto". This is called connecting with the audience and taking the audience to that stage where they feel as if they can visualise the context and hence "be a part of the song" being sung. Such is the power of using different senses in performance of different types.

FIGURE 3.1 Multisensory ambience creation for singer Leto. Photo by Wendy Wei: www.pexels.com/photo/man-holding-dj-mixer-machine-2350325/

3.3 AUDIENCE REACTION AND ATTENTION

It's very important for the performers to engage with the audience, in the absence of which a performance does not make any sense. In some performances like magic, circus, etc., the performers come close to the audience all in an attempt to connect with the audience. This mode of connecting closer to the audience and bringing the audience to the stage (in magic show) is to establish the bondage with the audience by reducing the distance between performers and the audience and also to establish a direct eye contact with them so that the level of bonding with the audience increases. It's easy to get the audience's reaction and also their attention after the performer has gone back to the stage. The mental model of the performer which is created remains in the mind of the audience and they get a feeling of closeness with the performer. Once this is achieved, the overall performance reaches a new level of satisfaction, appreciation and perfection. The audience then connect much better and the performers need not put any extra effort to draw their attention.

Magician "Duggi" is performing on stage. The auditorium is full, but she can hear murmur among the audience, a sign which tells her that they are not engaged with her and her performance. Duggi changes her performance strategy. She starts with the magic where she herself gets inside a box which is locked from outside by her assistant. Then a fire cracker tied to the box explodes, and the smoke engulfs the entire stage. The audience is tensed, the magician? Is she all right? Suddenly the spotlight from the stage turns towards the audience and magician "Duggi" is seen entering the auditorium and walking towards the stage. The audience is excited to find her here. There is a loud round of applause. The audience is seen to shake hands with "Duggi" and thus Duggi connects with the audience by reducing the distance (physical) with the audience and also reducing the hierarchy (stage) and coming to the audience so close. Then it becomes a truly magical performance with the audience deeply engrossed and expecting when would "Duggi" come closer to them once again. This is the power of proximity and reducing hierarchy in performance.

3.4 CAMOUFLAGE ON STAGE

In many performances it becomes necessary to hide or camouflage performers, props or certain elements. The need for doing this might be multiple. At times because of lack of time it is not possible to move heavy props, sometimes at the end of a scene it becomes difficult for performers to move very swiftly away from the stage. At times it might be that the scene being enacted demands that some parts of the scenes are highlighted and some parts are masked. The main aim is to ensure that the audience is forced to see what is being highlighted and not what they are not supposed to see. Many of us are familiar with that popular card trick. The magician shows the audience five cards and asks them to remember one. He then says that he would vanish the card that the audience have selected. Now he presents five cards and lol!! The identified card is gone!!. The trick is very simple. The audience is too engrossed in finding the card that they have identified and the magician takes advantage of this and shows them a complete new series of five cards!! Thus, the identified card vanishes.

A war scene is being enacted on stage. There are lot of killings. Those killed cannot leave the stage as it would look very odd. The duration of the scene is also very long, spanning over forty-five minutes. Those killed thus have no option but to lie down on the stage motionless which is very difficult. There are small insects and mosquitos as the performance is taking place in remote rural area. This is where camouflaging the dead solders becomes important. Either the area they are lying down could be made dark so that the audience is unable to see what's going on or the entire area could be covered (in synchrony with the scene) with a fabric which merges with the stage colour. In magical performances camouflage is used for hiding the performers, or for hiding a part of the performance to ensure that the trick is not revealed to the audience. In most of the performances the lighting on stage is made to flicker so that the audience is unable to see anything and this time is utilised to make necessary changes on stage. Flickering light causes disturbance in visual field and even after its removal the after-images in the eyes do not allow the audience to see clearly. This time is utilised for making changes, mask elements or prepare for the next surprise on the stage.

3.5 VOICE MODULATION IN ATTENTION GRABBING

You like to listen to the speeches of some people, whereas you feel bored while listening to the speeches of some. In performance as well, if performers are speaking in the same tone and trying to connect with the audience, then it becomes pathetic and boring to listen. This is where voice modulation becomes important. When you modulate your voice while speaking in any performance, a pattern is generated which is dynamic. It's easier for the audience to focus on what you are saying by focussing on this pattern. Humans are as such good at pattern recognition. Modulation involves giving pause, changing your pitch and emphasising on some words more compared to others. The audience visualise these patterns and it grabs their attention. So when you are speaking, talking or delivering your dialogue, voice modulation along with gestures plays an important role in getting the audience's attention. When performers present over the radio, this becomes still more challenging. One of the channels through which the audience receives the information is now absent, that is, the visual channel. So the entire onus goes on the voice modulation which is transmitted to the audience through the ears. For novice performers the audio performance with voice modulation becomes difficult while sitting in one place. So they are advised to make body movements and gestures assuming there are audience sitting in front of them. This makes voice modulation easy for novice performers in creating the proper context for better voice modulation.

Mr Pancho is a comedian. He performs on stage as well as on radio and seems to be well conversant with both the media. He was once presenting on stage and the place was in proximity to a busy railway station. When he started performing the audience started appreciating his performance. The problem started after an hour when the movement of the local trains started and their frequency increased. The vibration of the trains and the sound of the horn were now creating problems. In spite of having a good sound system, a large portion of the dialogues was not audible. The entire show was a flop and the organisers had to apologise to the audience and reschedule the show at another venue.

The above is a problem in many cases. If the programme venue is in a noisy place, then it will interfere with the performance. Here any amount of voice modulation would give very little results. The only way to overcome such a situation is to deliver the dialogues when there is no noise (which needs careful timing) and when the noise starts use body gestures to augment what has been delivered or is about to be delivered. An example would make this clear. Let's imagine that a comedy performance is going on at the same venue. Then one needs to know the frequency of the trains and their timings. On analysing the train timings and the gap in between two trains, the dialogues for the comedy could be written. Care has to be taken to ensure that in the interval in between two trains the dialogues would be delivered and when the train movements happen gestural communication linked with the dialogue should be performed. This is tough to perform and needs repeated practice at the venue. Thus, when your ears are overloaded with sound the focus in performance should go on gestural communication. As an alternative, performances should be arranged which need to be seen and not to be heard. For example, "mime" performances are based on gesture and there are no dialogues there. In case of gesture-based performance any dialogue looks or sounds weird.

3.6 ENSURE THAT IT IS NOT NOISE AND DISTRACTING

Young people have a habit of listening to loud music. For elderly or middle aged people it appears to be noise and not music. So for some it's music and for some the same thing is noise. This has to be taken care of. When there is any performance sound plays an important role. It should be adjusted to a level so that everyone in the audience could perceive it as music or it should appear pleasant to the ears and not appear to be noise. Always remember your audience should enjoy and it's a mixed population of different ages. Too loud sound is distracting and is perceived as noise by many. The details of any performance are lost when the sound is too loud. It's better to adjust it with users of different age groups sitting at different locations at the venue. You need to ensure that those sitting at the far end of the venue can hear but those sitting in front do not perceive it as noise as well. That's the reason why sound boxes need to be evenly distributed all over the venue, based on the audience distribution at the venue.

3.7 YOU NEED TO BE HEARD

Nipen is a very popular playback and stage singer. He mesmerises the audience wherever he goes. While he sings he jumps, dances and he even gets down from the stage and shakes hand with the audience and again goes on to the stage while singing. He is seen shaking hands while bending down from the stage with his audiences. He is a darling of the audience not only for his mesmerising voice, but also for his way of communicating and connecting with the audience physically. This makes the performance an engaging one where the audience also feels a part of the concert and it's no longer a one-way communication. Humans always prefer communicating with others in whatever mode it is. While singing, Nipen immerses himself within the crowd, the crowd feel that he is one very close to them. An automatic affinity is developed between Nipen and the crowd. While singing Nipen at times turns the microphone towards the crowd encouraging them to sing along with him. This keeps the audience alert as no one wants to miss the opportunity for singing.

For being a good performer you need to be "heard". This does not mean that you have to shout aloud, but it means that your presence should be felt by the audience and all that you deliver on stage as a performer should reach your audience. This is a combination of many factors such as your attire, the props, lighting on the stage, your way of delivering dialogues or music to the audience, etc. The biggest challenge is that you always do not get a direct feedback from your audience. At times they clap, cheer or communicate with gesture. At times they are silent. These are important for the performers to get to know whether they are able to connect with the audience. To get a feedback in case the audience is silent there are certain things that performers do to ensure feedback. A vocalist at times while singing comes down from the stage and starts walking among the crowd while continuing to sing. This helps them to see people's reaction and connection. At times the vocalist while singing on the stage turns the microphone towards the audience and asks them to sing. The audience sings a few lyrics and the performer gets a direct feedback from the audience that they are connected with him/her.

So whether the performer is being heard needs to be checked to ensure that the one-way communication between the performer and the audience is effective. If there are no reactions from the audience, then the show possibly is not going on well and the performers need to change their strategy of presentation. In group performance like theatres and movies there are at times central characters and the focus of the audience is primarily on them. A popular hero or heroine is always the centre of attraction. This does not mean that the supporting or co-stars should not be given any importance. To ensure that the group performance is equally emphasised on all the performers, it should be ensured that every actor tries to connect with the audience so that the connectivity between the actors and the audience is established. This is where the lighting on stage plays a role in highlighting all performers on the stage so as to establish coherence in the performance and not create a hierarchy among the performers. This shows that audience feedback to performers is the key to a good performance.

3.8 HUMAN FACTORS AS A TOOL FOR ENHANCING COMMUNICATION

Today the world is heavily dependent on communication. Performing arts is a powerful medium of communication, wherein a message is conveyed to the audience. The content of the message, its length and form of transmission vary and are mostly geared towards the audience's expectation and the current geopolitical situation in existence. This communication can be enhanced by following some simple human factor rules. One of the many challenges that performers face is keeping the audience engaged. If the audience is bored and does not engage with the performance, then the show becomes a flop. So how to address this issue?

Your audience seeks variety in performance. This variety or change on stage is important to get the audience connected with you as a performer. A magical performance is able to connect better because there is constant change and newness at every stage and thus the audience gets connected. In a classical musical performance where there are minimal body movements or gestures your audience can get bored. The only way out of this (if possible) is to engage the audience in your performance either through claps or to sing a few lyrics along with you as the performer. The second alternative would be to introduce variety in your classical songs. It's possible that as a performer you sing classical music of a specific genre. To bring in variety, change the genre after some seven songs. This brings variety. In between songs cracking a few jokes with your audience or taking your audience requests for singing a particular song is another way. These small acts connect you with the audience, breaking the set monotonous pattern and creating a varied pattern.

In case of theatres, opera, ballet, dance, etc., the same element of variety needs to be incorporated. In this case human factor principles of using multisensory elements like sound and light in varying amounts can create a specific sequence, which connects with the audience. Dialogue delivery should ensure that there is a change in the pitch and it is not monotonous to draw the audience's attention. While using colours refrain from using pure green and pure red and add impure variety of the same. This is because a large spectrum of the population is red and green colour-blind and would visualise these colours in shades of grey.

3.9 RECORDING AND PLAYING

Dhanu Hondon is a new singer. He has become popular all of a sudden after her song became viral on the social media platform. Prior to this, he was singing and begging on the streets. His song was seen and heard by a famous music director who took him to the town "Aram" for playback singing. For Dhanu this was a new experience. As he was taken inside the recording studio, and made to stand in front of the microphone, he started feeling suffocating and uncomfortable (Figure 3.2). When he was asked to sing he failed to do so because the ambience was completely new to him. He was used to sing in the open as a beggar. That spontaneity in his singing was completely gone. The music director took him to the open garden and asked him to sing. Now he could sing very well and it was a different Dhanu singing now. The director asked Dhanu "what's the problem?" Why are you not able to sing inside the studio? Dhanu replied: Sir I feel suffocated, and feel detached from the world outside, how will I sing?

Recording of voice (dubbing), songs or music in controlled environment is an important part of performing arts today. The recording rooms should set the right mood for the performers, else the performance would not be optimal. The

FIGURE 3.2 Dhanu singing inside the recording studio. Photo by cottonbro studio: www.pexels.com/photo/back-view-of-a-woman-recording-a-song-in-a-music-studio-7086742/

size of the room should factor in the body dimensions of the performers and be designed accordingly. It should be considered whether performers would sit or stand as sitting would occupy more space than standing.

You have to remember that while recording in studio at times to bring in that emotion or mood in the voice performers might need to move their body parts or even enact the scene (though not necessary). This demands additional space based on body dimensions and body movements, which needs to be provided. For example, a performer who has to dub the dialogue for a fighting scene might need to enact the same to deliver the dialogue.

In the dubbing room, the biggest challenge is to synchronise the dialogues with the lip movements. People have a limited capacity of their short-term memory, and they often forget the context of the dialogue they delivered while shooting the sequence. At times they fail to bring in that emotion while sitting down and dubbing. For example, a dialogue in a fighting sequence is difficult to dub while sitting still on a chair. Some point of reference needs to be built where the sequence is played and the original dialogue delivered on the spot (which is not kept, because of stray noise) is also played simultaneously. This is to help the artist retrieve the actual dialogue style and deliver it with requisite emotion and feelings.

Singers while recording also need to move or move their body parts. So the space should account for this. In an enclosed room often people feel claustrophobic or suffocated. To remove this there should be a connection to the outside world through a glass window or door which might be closed. If this is not possible, then mirrors could be attached to make the space look bigger. In case this is also not possible, light colours could be used to make the space look bigger. As an alternate the performer could be kept engaged with the performance the moment he/she enters the room. If they focus on the performance, then the feeling of claustrophobia would go away. Another way of dealing with this situation is to ensure that all the small rooms should be provided with glass walls so that the users are able to see and "visually" connect with one another. This also removes claustrophobia to a great extent.

Sound and music recording demands that the performers sit or stand and perform in front of a microphone which is at a fixed height. This is where the height of the performer's lips has to be accounted for and the microphone place accordingly. Mr Kanu Mondol goes to the recording studio to record a song. He stands in front of the microphone only to find that the height of the microphone is too low and he has to stop forward. He started getting back pain and after sometime is unable to focus his attention on the song. He then decides to stop and informs the music director of the problem. Thus, height adjustment of microphones as per the lip height of the performers is very important in enhancing the performance, especially in recording studios.

3.10 DEVELOPING YOUR STORYLINE IN TANDEM WITH MENTAL MODEL

In film, video, theatre, etc., the storyline is the heart and soul of the entire performance. The audience go to watch such performances as they see themselves in the performance or visualise the events showcased in their day-to-day life. This glues them to the performance. Hence, the storyline should be designed in tandem with this mental model or picture engraved in their brain. You need to remember that users are very bad in short-term memory and cannot keep lot of information in their heads at one go. This is very pertinent when there are too many characters in a scene and the audience is unable to keep a track on all of them. For elderly audience the problem of remembering so many characters and their names and roles becomes all more difficult. The storyline or the script should ensure that it has some landmarks which could be some song sequence, twist in the storyline deviating from the norm or some particular place which people could relate later. Always the storyline cannot be in tandem with the audience's mental model, else the surprise element in the film or theatre would be gone.

The attention span of audience is very low and to glue them to the performance it's necessary that there are changes in the storyline and lots of surprises. The incorporation of multisensory ambience through the prudent usage of sound, light, and costumes in varying proportions could enhance the audience's interest. If necessary intervals should be given for the performers to rest and the audience can also take a break and then resume further, in case of live performances.

"Leto" is a script writer for theatres and cinemas. He picks up good novel and converts them to scripts for performance. He gets a new assignment for writing a script for a theatrical performance. He picks up a novel written by "Duggi" which is based on the story of a poor rural girl who struggles to study in her village and then goes to the city for higher education. Leto looked at the story and started scripting it as he felt that the audience would like it. First of all, from a human factors perspective the story is very close to what the society is in reality with struggles for rural children when it comes to education. This is true for the location of "Eldorado" where the context and the audience belong to. Leto started crafting the script for the theatre very carefully. He introduced some musical and dance sequels for breaking the audience monotony and entertaining them. There were some laughter and romantic scenes added as per the storyline so that the central theme of the story which is "struggle of a rural girl in higher education" remained intact but variety was added to ensure that the audience is not bored or fatigued. When the theatre was staged it was a bumper heat.

In another occasion Leto received a contract for writing the script for the same story but this time for a movie or cinema. He used the same process but improvised with adding different locations like the hill, sea shore, and desert to add to variety and make the sequel interesting. He ensured that the dance and musical sequels were points of reference (landmarks) with reference to which the audience would be able to recollect the storyline and remember the sequence of activities. He preferred to give the movie a vintage look and hence it was in black and white and the costumes were chosen accordingly to generate a mental model among the audience of the old era. The costumes were also chosen accordingly. The musical sequences were crafted carefully to include only sounds of vintage musical instruments to create that old era among the audience. At the end when the casting was shown it was done in colour to bring the audience back to the present and remind them of their trip down memory lane on a time machine. The movie was another superhit among the audience. It was a mixture of fantasy and reality which was well balanced.

3.11 MAKING THE SPACE RELEVANT TO YOUR STORYLINE

A street play is being staged at the entry to a busy market. The sequel being staged is the hero and the heroine eating ice cream and slowly walking down the sea shore. They are at times sitting down facing the sea and the heroine is throwing pebbles into the sea. The audience is confused as to what the play is all about. It's hard to imagine a sea shore at a busy market! One has to follow the dialogues carefully to get to know that it's the sea shore, else the total play is a chaos.

When you are performing on the street or on the roadside, the theme of the play should match the roadside. The romantic scene with the hero and heroine singing on the sea shore would look odd if performed on the street or the roadside. For such locations the script and the story line have to match the sequence so as to be in tandem with audience expectation. It's difficult for people to imagine in the context a sequel which is related to the sea shore or the mountains. It then becomes confusing for the audience to relate them to their daily life.

Similarly theatrical performances in controlled environment in halls are relatively better as here we can create some space in line with the storyline. For example, bedroom, living room, and classroom could be easily created with the help of props. Thus, it gels well with the storyline and the performance is augmented. For movies or cinemas we have lot of freedom of shooting in the location relevant to the script. This is the best option for building the storyline and the script coveys the story to the audience in a realistic manner that they are able to relate to as per the script or the story. In such cases the audience also remain with the movie and travel with it to different places and sequences.

3.12 JUDICIOUS USAGE OF COLOURS

Roga and his girlfriend Mota go to see a dance drama being performed in town. The troupe performing is a famous one and getting a ticket for the event is difficult. Roga is excited to see the performance and so is her friend Mota. They take their seats in the auditorium and the show starts. Both the friends are spellbound by the troupe's performance. Suddenly Roga tells Mota, "see, I feel they should have worked on the costumes". The costumes in grey were a little odd for the performance (Figure 3.3), Roga says. Mota is surprised, and says

FIGURE 3.3 Colour-blind audience unable to detect red and green in performance. Photo by Jah Nomad: www.pexels.com/photo/ballerina-in-a-red-costume-17835679/

"where did you see grey costumes?" They were all of bright colours mainly in shades of red and green. What happened was that Roga was a colour-blind person, to be specific red and green colour-blind. This means that he cannot see pure red or pure green but sees them in shades of grey. Interestingly only males are affected by colour-blindness. His girlfriend was normal sighted, and she could see the colours red and green and thus was surprised when his boyfriend criticised the costumes for their dull grey colours.

Performance of any type should be colourful because your audience loves colours. We need to be careful with just one aspect of colour that a large number of the population are pure green colour-blind. A few are pure blue colour-blind. So what this means is that these audiences would see pure red, green or blue in shades of grey. It's difficult for you to know who is colour-blind in the audience. So the best option would be not to use pure red, green or blue colour. You might add some impurities to these colours and make them impure and that would work fine for all. For example, a little white or black in green or a little white or black in red retains the colour but slightly changes its purity. While using coloured dresses, props, lights, etc., this could be followed for making your performance acceptable to all.

Performing arts is a place where a variety of colours could be used for setting the mood of the performance. If it's a rock concert, then abundant usage of colours in the form of light, costume, and props is normal and gels with the mood of the performance. On the other hand, a classical musical performance does not demand much colour usage which becomes distracting as the focus of the audience is on the music and not on the colours which add to distraction. Even the performers wear simple, sober dresses which are not colourful as the mood demands soberness. In theatrical performance colour has to be used for setting the mood of the scene and developing the mental model or the image of the audience. In a birthday party there has to be use of lots of colour. In a classroom scene the colour usage has to be restricted to two or a maximum of three, else it would be distracting the audience from the conversation and the act being performed.

For street performances colours play a role in getting people's attention. This is because on streets there are lots of distractions, and it's essential to draw public attention to the performance. One of the ways of doing this is through sound and music. To augment these further colours can be used. This could be used in the form of coloured costumes, props, etc. These sound and colours help in drawing people's attention and help the performance stand out from the crowd and the noise on a busy street.

Coloured lights are an integral part of any performance. These are used in musical performances where the change in colour is associated with the change in performances mood. In theatres coloured lights set the tone of the particular scene. A blue light on a dark stage sets the tone for moonlight night. A red light on stage is used for depicting violence. Dance performances involve change of lights of different colours. As the audience has an association of colour with different events, coloured lights help in creating that mental model (mental image) of the scene to the audience and thus connecting with the audience becomes easier by the performers.

3.13 HOW TO SHIELD ARTISTS FROM DISTRACTION IN CROWD WHILE RECORDING

Ghantu is the heroine of a film "Kukurer Nam Leo". The shooting for the sequel that is being shot is that she is chasing a pickpocket who has just stolen her purse from her side bag. She is running through the narrow and busy streets of "Faltu" and the crowd is looking at her in surprise. Many times the crowd is coming in the way and the shooting staffs are requesting the public to stay away from her path so that the shooting goes on smoothly. It's becoming very difficult to control the curiosity of the crowd. The shooting at times is also being affected. The director seeing this addresses the crowd. He explains to them the film for which the shooting is going on and asks them to cooperate so that the shooting gets over peacefully. He tells the crowd that if they cooperate the movie will highlight their place and the place would become famous thereafter. This worked and the crowd became organised and the shooting was completed without any disturbances.

Film artists face lot of challenges while shooting outdoors. The crowd gather and start looking at the sequence and they have to act in this context. This distraction can at times affect the artist's performance and lead to problems. One of the ways to shield artists from such distractions is to shoot sequence which has lots of actions instead of only dialogues. For example, fighting sequences are best enacted outdoors because the movements are so fast and demand focus on specific body parts that the actors do not get to focus on the crowd and their movements. This is because to perform such actions there has to be coordination between different body parts, which demands concentration on the task. Because of this they are less distracted by the crowd around. Another way to shield artists from distractions from the surrounding crowd (without driving away the crowd) is to involve the crowd in the shooting process (Figure 3.4). The crowd could be briefed that they are being filmed as bystanders, but they need to be briefed that they have to behave properly and not make a chaos. This works as everyone has a hidden desire of acting in a film. Even if they are not included in the shoot, the crowd could be told this to ensure that they are organised and the shooting takes place peacefully without any distraction to the artists. Human being wants some sort of recognition and identity. If they are promised the same, they become organised and behave well.

FIGURE 3.4 Involving the crowd in film shooting. Photo by Mathias Reding: www.pexels.com/photo/crowd-on-street-market-in-old-city-4397894/

3.14 KEY POINTS

I. All the senses of the body should be used for creating a multi-sensory ambience.
II. It's important to get the audience reaction for performers to ensure best performance.
III. For creating camouflage on stage for different purposes light plays an important role.
IV. Performers need to be heard and voice modulation is one of the ways to ensure that.
V. Care should to be taken to ensure that the sound doesn't turn to be a noise.
VI. Feedback from the audience is the key to ensure an engaging performance.
VII. To ensure audience feedback performers have to leave their usual position on stage and come in proximity to the audience.

VIII. For musical performance involve your audience in the performance and make them sing a few lyrics. This makes the performance better and the audience gets involved.
IX. Variety is the key in performance for keeping the audience engaged.
X. In studio environment communication between different units or with the outside world helps in eliminating claustrophobia in artists.
XI. The storyline in performance should be in tandem with the audience's mental model. To ensure this use landmarks, music, dance, etc., so that users get connected and are able to recall the sequel with reference to these points of reference or landmarks in performance.
XII. The context or space in performance should match the storyline.
XIII. Use colour judiciously and factor in colour-blind people.
XIV. For shielding actors from distraction of crowd either shoot only action sequel or involve the crowd in shooting.

3.15 PRACTICE SESSION

i. A magical performance has been organised in the city auditorium. The auditorium is very old. Your expertise as a human factors specialist has been solicited for ensuring that the performance goes on well. The magician and his team have also approached you for your advice. What human factors advice would you give to the magician and his team as well as to the management of the auditorium?

Human Factors Directions

A. Look for the illumination and the sound system to ensure that they illuminate the stage and the sound system is placed all over the auditorium.
B. Discuss with the magician whether for some scene the sensation of smell could be used.
C. For those scenes requiring a part of the props to be hidden plan for specific lighting instead of making the stage dark. Flickering lights might be thought of.
D. Check for the visual cone of the audience from every corner of the auditorium.
E. Check for smell sensation from every corner of the auditorium.
F. Ensure that any camouflage with props works from all angles of the audience.

ii. A comedy performance has been arranged in the open. The venue is a big filed which is very close to the railway station. The problem is that the sound of the trains can create problems in any performance in the vicinity. What human factors issues would you take cognisance of for this performance?

Human Factors Directions

A. Check for the timings for different trains.
B. Discuss with the comedian the issue of noise of the trains.
C. Tell the comedian about the exact time interval where there are no trains.
D. Sit with the comedian to restructure the comedy script ensuring no dialogue is delivered when the trains pass.
E. The noisy period could be filled with gesture supplementing the dialogues.
F. To mask the noise background slow music might be used.

iii. A circus has come to the town. They have a number of performances and the trapeze act is the famous of all. They have approached you as a human factors expert to help them create an ambience for this particular performance so that it becomes the prime attraction of the circus.

Human Factors Directions

A. Study the nature of the trapeze.
B. Note down the duration of the trapeze act.
C. Choose appropriate music which synchronises and keeps the tempo of the performance. For critical balancing act you might include suspense music.
D. Use lighting that focuses on the entire performance as well as on the specific performers.
E. At the end finish it off with nice farewell music and bow to the audience.

iv. A band is playing in front of a shopping mall. It's night-time and there are lots of crowd and traffic in the vicinity. The band members are finding it tough to get the audience's attention and it seems that people are just ignoring them. As an expert what human factors approach would you suggest so that the band gets some audience?

Human Factors Directions

A. Make the band members dress in fanciful colours to stand apart from the crowd.

B. The group should not only play and sing at one place but also move. The dynamicity will attract crowds.
C. If possible the musical instruments could be wrapped up in fluorescent colours to draw public attention. For example, the sling of the Spanish guitar could be of a fluorescent orange colour.
D. The band could move and shift their position after ever song if possible – for example, once on the staircase, then the pavement and so on.

v. A would-be singer is recording her first song in the studio. She is feeling claustrophobic because the studio is too compact and she is only surrounded by walls. She is unable to perform and is going for repeated re-recordings. Your human factors expertise is solicited here.

Human Factors Directions

A. Provision for a glass window if possible for getting a view outside.
B. Having a glass panel on all sides through which she could look around.
C. Some large paintings, for example, scenery on one side for breaking monotony and reduce claustrophobia (sense of suffocation within an enclosed space).

vi. A script writer needs your help as a human factors expert in writing a script based on a novel which is all about a travel journey of a school boy from home around the world and back to home without informing anyone. What human factors advice would you give him for this script which would go for making a movie for children?

Human Factors Directions

A. As this is for children costumes should be colourful.
B. Less characters with more scenery.
C. Humour to be included and no serious dialogues.
D. Dialogues to be short.
E. Music and song sequence should be augmented by movement.
F. Use animals (real or props).
G. Give an overview of the story before the film starts. It prepares the young audience what they are about to see.

Editing and Scripting and the Challenges of Street Performances

4

4.1 OVERVIEW

This chapter gives the readers an insight into the different human factors issues related to editing and scripting for movies and theatres. The readers are taught on the application of human factors related to selecting the appropriate fonts and colours. The human factors aspects of deciding upon the background and foreground colours are discussed here. The chapter talks about the judicious usage of colour while editing. The best ways of eliciting audience emotions through prudent script and editing are also discussed. The different human factors directions for writing a prudent script that would make an impact on the audience are presented. The chapter takes the readers through the human factors attributes of crowd gathering in performances. It explains the ways and means of effective communication to ensure successful performance. The readers are given clues on handling nuisance on the streets while performing and how to eliminate noise or ignore it while performing and creating a different ambience without making any changes to the surroundings.

DOI:10.1201/9781003506409-4

4.2 CHOOSING THE RIGHT FONT FAMILY AND TYPE

Textual information plays an important role in movies and even theatres where specific information needs to be communicated to the audience. In theatres the need for them is less compared to movies but still it's used. In theatres textual information is used in some scenes to convey name plates at office doors, signage inside a building, protesting people with placards, etc. But in movies text is used for depicting the names of different people associated with the movie and at times also supplements specific movie scenes like "once upon a time in 'Telebhaja' there lived …" (Figure 4.1). While choosing text you need to be careful in choosing those font family with which your audience is familiar. You might test this out by talking to a few people. It's important that a familiar font is used, else the audience would have problem in decoding them. They might also make mistakes in identifying familiar words. For example, if the audience is familiar with the font family "Times New Roman", then it's wise to use this font for conveying any information. For example, if you know a person for long, then whenever you see that person in different places you tend to recognise the person easily and do not take time for the same even though the place has changed. Here also similar things are applicable.

The next challenge is whether one should use uppercase and lowercase or all capital letters in conveying textual information. Upper- and lowercase when used is a better choice because it generates a pattern. If you carefully look at a text in upper- and lowercase, you would notice that because of the unequal shape of each letter you would be able to trace a pattern around the word. This pattern varies with different words. When your audience read the text from a distance, they would not read them alphabet by alphabet but read the word as a

FIGURE 4.1 Movie captions setting the storyline and preparing the audience for what they are about to see.

complete pattern. This makes the reading fast and accurate from a distance. If you use all capital letters, the shape and size of all the letters in a word are the same and they do not generate a pattern. The users thus have to read the words individually which takes time. But when you need to draw the attention of your users, all capital letters are used as they tend to "shout" and stand apart from the crowd of information. Thus, we use all capital when we want to draw attention and upper- and lowercase letters when we want to read with ease and fast.

4.3 MOVEMENT OF TEXT FOR BEGINNING AND END

Subtitles are quite common in movies when it comes to a movie in a language that the audience is not familiar with. The subtitles are always better placed just below the images. The reason is that the human eye scans from top to bottom. So the eyes first see the image from the head to the foot and then the subtitle is read, though the process takes place very fast. Here also the combination of upper- and lowercase is preferred. For the subtitles to appear there are two ways: one is that they appear as the dialogue is delivered and appear below the character. It becomes easy for the audience to relate. The second way is when subtitles or texts need to appear on the screen for conveying some important information like "1942 a remote village in Western Eldorado"..., then it's better if the text appears from left to right because that is how the majority of us read. In places where the audience reads from right to left, the appearance for such a text should be from right to left.

When a movie starts the main characters, for example, are casted. It's like "Duggi Leto in and as Pagla Kamre De" ...Ideally this should appear at the central part of the screen as that is the most prominent part of the visual field. This text can either appear at one go or appear from the bottom upwards. The appearing of the casts in movies from bottom to top is preferred because humans "scan" or read from top to bottom (Figure 4.2). Thus, if the casting text moves from bottom to the top and your eyes scan from top to bottom, easy reading is facilitated. It's like two friends are moving towards one another. The faster they move, the quicker they meet and interact. Similarly as in casting not much time is available as the movement from bottom towards the top facilitates quick and easy meeting of the text and the human eyes and thus facilitates faster and easy reading. In many movies the names of the hero and the heroine are displayed at the end. Here it's better to use the central part of the screen to show prominence and importance to the audience. In all these cases the impact is better if accompanied by appropriate background music along with dynamic text which changes its font size, becomes bold or changes colour.

FIGURE 4.2 Casting text movement in films.

4.4 BEST WAY TO DEPICT EMOTION

Emotion is an integral part of your audience. To evoke emotion you have to play with the script and use appropriate dialogue that stimulates the same. For theatres which are live performances tears and laughters are used in scripts to evoke emotion. These two are contagious and spread among the audience very fast to evoke emotion among them. You might see that when one cries the person next to the one also starts crying or moves into that mode. Similarly when someone laughs in the crowd it rapidly spreads among the full crowd. This is the impact of stage and movies on the audience in evoking emotions.

While scripting for theatres and movies the script should weave emotion in the dialogue itself. The aspect of emotion has to be looked into from a broader perspective of the audience and what their needs and wants are. When your audience go to watch any performance, either theatre or movie, they like to see themselves in the storyline. If the storyline and the emotional content like laughter, sorrow and tears match their emotions, the performance connects with the audience and becomes popular. The emotion can also be generated if the needs of the audience are mapped into the script. This could be achieved by talking to the users about what elements evoke emotion in them. For example, for some it could be an old song, a hill station where someone had been in childhood, school, grandparents, etc. These range of options need to be studied and incorporated into the storyline and the script to create that framework for eliciting emotion among the audience. The issue here is to find a common emotional "landmark", that is, the incident that would elicit emotion for the entire spectrum of the audience.

Nipen Da is the director, script writer and novelist for a new movie "Bansh". He is looking for some emotional content to be included in the script and prefers to do some user study. He travels the length and breadth of the country and gets some data related to the elements which trigger emotion. His list is quite large and he has to choose those common emotional points that would elicit emotion for all. Finally he narrows down to fifteen emotional elements for his three-hour movie. Some of the elements that he found were elderly parents, childhood best friends, games played in childhood, first trip with family and pets. He uses these "landmarks" in his novel and script for making his first film. In addition to this, he also does a test run with a few users before the film is released. This is necessary to ensure that the emotional landmarks are correctly selected and placed in the script. While delivering the script the actors need to modulate their voice accordingly and this should be accompanied by music and light which would further set the mood for the same. Thus, they try to create a framework within the script for eliciting emotion.

4.5 BACKGROUND AND FOREGROUND SELECTION

Performance whether live or in cinema is a collage of many different elements. When an artist performs on stage, the background design and colour are important as they help in forming a collage of the performer. If the performance is singing or music, then the background should change in tandem with the music. The audience gets an enriched experience when the background is set according to the performance. If the budget is low, then we recommend that keep the background on the stage of a colour which does not distract the audience and helps them to focus on the performance. For example, a light sky blue with some dark text or design or deep navy blue with some light text or design would be a better choice. In case the background theme has to be changed, then the change should happen in accordance with the music. If a song is all about children having ice cream on streets, then the same should be shown on the background to help the audience develop an image of what the song is all about and they are then able to relate and connect themselves to the same. In case of movies the background has to be real as per the context, that is, for those involving the sea shore the same should be there in the background for theatres and movies and if possible for musical performance. For magical shows the background should also match with the theme of the performance. This helps the audience generate a mental model and prepare for the show and get themselves immersed into the performance. At times pure musical performances or vocals are supplemented with dance on the stage.

4.6 JUDICIOUS USAGE OF COLOUR

The usage of colour in live performance has been discussed before. In film and video the colour usage has to be done judiciously. A red coloured car or refrigerator is a preferred choice but someone in a red shoe in office looks funny! Depending on the culture of the place, colours should be used in performances. When food is shown in movies the audience get a flavour of the same through colours, which are the look good factors that make your audience hungry. Similarly, for violence scenes red colour, which depicts blood, generates the feel of violence because it is associated with blood. A green ambience in violence would look weird. For textual information provided on different sequences in movies and video while editing it should be considered that the text pops out from the background. Thus, if you are using text in the backdrop of a forest, refrain from using green and use golden yellow instead. Similarly, if deep sea is in the background, then refrain from using light blue and use deep navy blue instead, which pops out from the light blue colour of the sea and also gels with the theme of the background. The users are able to relate them sitting in the auditorium.

4.7 HOW AND WHEN TO POSITION WARNINGS

Warnings in movies are a part of government regulations. These need to be provided in a manner such that the audience could understand and would be able to decode with ease. For example, warning related to violence should be given in the central part of the movie frame that is the most prominent part and the eyes also see it with its centre aligned for maximum clarity. All warnings should be static at least for a couple of minutes on the screen and should stand apart from other writings on the frame. Warning should not be written in any stylish manner and should be as simple and concise as possible for ease of reading. As the auditorium is dark, a dark background with light text (provided that it's a one or two lines of text and not more) provides the best readability for audience. The dark background makes the pupil of the eyes to dilate and the white text reflects light. For smoking and alcohol-related warnings also the same principle should be followed. The text should always be accompanied with voice-over where the warning should be read slowly but firmly, preferably in a heavy voice of male so that it's better registered with the audience. All warnings should come before the frame so that decision-making becomes easier and the audience is ready and alert. When warnings are displayed in movies there should not be any background music or sound to increase attention and show the gravity of the situation. In live performances warnings or safety instructions related to emergency exit should be given in bold and loud voice. To begin with a bang sound could be used to draw the audience's attention and following that the announcement could be made for their attention. The loud bang sound is in tandem with the mental model of the audience. If possible such warning messages when communicated in live performances should be repeated in different languages depending on the audience present.

4.8 SCRIPT WRITING FOR EASY RECALL

A good script leads to a good performance in case of cinemas and theatres. Any script should first prepare the audience for the bigger story. This preparation could be in many different forms. This sets the stage for the main story to flow. As the audience have a very bad short-term memory (similar to the computer RAM), they need to be presented bits of the main story and in steps. These steps in the storyline should be logical and connected so that the audience is able to follow it till the end. Thus, scripts should not have long dialogues as the audience would tend to forget them. Short dialogues make more impact. Dialogues should be buffered with humour, music, dance or similar sequence to break the audience monotony and help the audience retrieve any sequence and dialogues with specific reference to the music or dance sequel. It is not necessary that all scripts for performances should be in the form of a story which goes in a sequence; it all depends on the novel selected for the purpose. The audience should be able to remember the characters and the sequence and for that reason a particular scene should not last long with too many dialogues. Too many dialogues in one scene overload the audience and they tend to forget the theme of the scene depicted through dialogues. It's better to use words in the dialogues which are short and not too long as elderly audience have problems in comprehending them apart from recalling them later. If the audience are impressed with some song or some dialogues in the movie, then they register them very well and the next day they are vividly able to sing the full song or articulate the entire dialogue.

4.9 CREATING AN IMPACT AND COMMUNICATION

Creating an impact is the biggest challenge in any street performances. There is a big challenge in effectively converting the street into a performance venue with limited resources. The key to drawing and retaining audience is effective communication. That's why you see that before any street performance starts the performers use their public address system to talk, shout or even sing. Some even dance to draw public attention. A street musician starts with his or her music, preceded by just instruments if he/she is a vocalist. At times, bright coloured banners also help to create the ambience and help attract the audience. Choosing the right place also plays a big role in communicating to the audience. Corners, dead ends, in front of prominent landmarks like buildings, statues, etc., are good locations for street performance. People get to relate the

FIGURE 4.3 Human factors aspects of performance during festivals. Photo by Christina Petsos: www.pexels.com/photo/men-and-women-in-traditional-clothing-singing-by-christmas-tree-19712205/

place with reference to the landmarks; thus later on for repeat visitors finding out the performers becomes easy with reference to the landmark. In street performance connecting with people on the move is the biggest challenge. People are only going to stop to see or listen to you if you present something unique which is not available elsewhere. To do this performers have to use the ambience around and gel the performance in tandem with the ambience to create a unique mental model. For example, if it's raining heavily, then a song related to rain gels well with the ambience and people on their way home or if they have time while going to their offices would stop and listen. This is because the song and the ambience are easy to relate and people can experience the performance and not just listen to it. This creates an impact on people and the performance is communicated well. Similarly, during Christmas time (Figure 4.3) any song of Christmas on the roadside would be appreciated by the public going home just before the Christmas break. This way of creating ambience in the middle of the street, roads or pathways breaks the monotony of the public and refreshes them with something new. Humans want variety in their daily life and these performances in actual setting and at the right locations can create impact and communicate very strongly. The target audience in such cases are those hard-working people who struggle daily to work to run their families. They do not have adequate funds for entertainments, and thus street performances become their only source of entertainment on their way to or from work. This mental framework of the audience has to be kept in mind.

4.10 CROWD CALLING

At times calling the crowd to a performance is needed. This is important where a dedicated space is allocated on the street or on the roadside for a number of performers. The public is confused as to which performance to attend. Thus, the performers need to draw the attention of the crowd for coming to their performance (Figure 4.4). For example, if the performance is all about balancing act on a tight rope, then the crowd's attention needs to be drawn towards that. In this way the crowd gets to know what the performance is all about and need not waste their time on something which does not interest them. Similarly, a performance which is all about playing a specific musical instrument accompanied by dance needs some articulation to the crowd to be aware of what it's all about so that they can take a decision depending on their likings. The information about the performance should be packaged in brief before advertising it to the audience who have a very short attention span.

The calling mechanism can vary. Simply shouting or asking the crowd to come closer to your performance might not work. In some cases it's necessary for

FIGURE 4.4 Crowd calling before performance. Photo by Darya Sannikova: www.pexels.com/photo/two-men-performs-on-stage-while-holding-rainbow-colored-umbrellas-2422474/

the performer to approach the public and convince them to attend the performance. Just mentioning the key features of the performance and explicitly telling them that it's not mandatory to pay to see the performance make the audience comfortable that they are not under any obligation to pay. In majority of the cases the audience would always pay for the efforts being put in by the performers. When a few people in the crowd pay, the others also follow the same and start paying them as well. Crowd calling is an art and one needs to play with the customers' psychology. Do not pressurise people but instead request them with the main features of the performance and speak in a tone that excites them. If you sell yourself in the performance attire, it becomes more realistic and people become interested in what you are about to perform. It creates surprise for your audience. Calling the crowd can start with a song, a loud bang or some dynamic activities.

4.11 SPECIAL EFFECTS ON STREETS

Special effects are used for creating the ambience for any performance. In auditoriums these are done through light and sound. In the open and in case of roadside shows creating special effects for ambience is difficult. When the known place is converted to a performance venue the mental model or image is developed faster and the audience appreciate or settle down much faster and in a better way. This known place then creates an ambience within which the audience are able to relate. Performance in such a context leads to an experience which is triggered through the eyes, the ears and even through touch and feel. This is where the real enjoyment from performance comes. In the daytime it's impossible to use lights for special effects. In such cases sound and music could be used for creating special effects – for example, a magician performing (Figure 4.5) during broad daylight on the roadside, demonstrating a "floating

FIGURE 4.5 Magic show on the street in the open. Photo by Roman Tymochko: www.pexels.com/photo/a-magician-blowing-big-bubbles-in-an-event-9315591/

person" . The ambience for this can be created by using musical instrument as well as by a loud narration. It might sound like this: "The floating man…in broad daylight…ladies and gentlemen please mind your heads there is a man floating above your heads…. This could be accompanied by suspense music. The same thing at night is relatively easier. Here flashing lights of different colours or a specific spot light could be used to create special effects. The colourful and light-reflective bright dresses of the performers can also aid in creating ambience in street performances. Some permitted domesticated animals like dogs, cats, etc., are also used in creating ambience, and though they do not directly add to the special effects they could augment special effects on street performance. Reflectors and paints are used in street performance to create some special effects in performances. For example, performances on micky mouse in broad daylight could have big motifs of the Micky and Donald painted on pavements which can go a long way in creating special effects in related performances. Ghost stories performed on the streets at times use big props which are huge masks of ghosts with big teeth, big red eyes and big pointed ears and these are of scale-up props and at times the scale is twice or thrice that of the original size. This also adds to the special effects for the performance.

4.12 CHANCES OF A NUISANCE

Any street performance has a chance of being vandalised due to public outrage. The public outrage happens if the performance hurts people's sentiments, which could be of different nature. Thus, to prevent such acts it's advisable to go for performances which do not have any religious, political or any antagonistic flavour. The best neutral performances which can be performed safely on the streets are magic shows, jugglery, health awareness, etc. Musical and instrumental performances are also good provided that they are neutral and do not tend to convey any message which hurts the sentiments of the people. Though street plays have been popular for conveying political messages to the general public in an easy-to-understand format, they come with the risk of agitating a section of the political parties who perceive these as a threat to them. The same holds true for musical performance. If the environment is conducive, then they could be performed for conveying political messages very effectively as music or rhythm goes very well with all sections of the population. They create a long-lasting impact.

4.13 NOISE

Noise is the biggest distraction in any street performance. This could be from people, traffic or a combination of all. You cannot expect everything to be silent on busy streets. So the selection of performance is very important to counter noise. Performances on the street should be selected carefully, and intricate, long and soft dialogues should be avoided. For masking the background noise slow music could be used, which supresses the distraction due to the noise and helps the audience concentrate on the performance which is going on the street. At times the noise also comes from other performances in the vicinity and can cause distraction to the audience and performers performing next to it. This is where placement of two road performances need to be planned. If a person is singing on his own playing the guitar, then a band should not play next to it. You might have a small jugglery show in the vicinity. Similarly, if a band is performing, then possibly a magic show in the vicinity would not be a bad idea as long as the magician is heard. The area covered by different performers needs to be calculated. For this their body dimensions and their movements along with different props and equipment need to be factored in. A roadside artist drawing someone's portrait could be sitting in proximity to a person singing on his own without any musical instrument. The song would not cause any disturbance to the artist and the singer would not be disturbed by the artist. At times the slow rhythmic music on the streets provides a soothing effect to the tired pedestrians walking along that path. In fact both could complement each other's performance. The area could be perceived as a makeshift performing hub, giving the public ample choice for prudent decision-making of what to watch and what not to.

4.14 KEY POINTS

I. Select upper- and lowercase letters to enhance readability and all capital letters to emphasise or convey important information.
II. Movement of text should follow audience expectations.
III. Emotion in performance can be augmented through music, lighting and sequence.
IV. The background and foreground should be calibrated to ensure it adheres to people's mental image or mental model.

V. Colour should be used in texts factoring in colour-blind people and the background.
VI. Warnings in performance should be displayed without background music and distraction.
VII. Script should be of short sentences and to break monotony it should be mixed with dance, music and humour.
VIII. Audience distraction is the key problem in street performances, with the audience having the flexibility of choosing from a wide range of options.
IX. Creating an impact through performance is the biggest challenge in street performance and this is where communicating with the audience through different modalities like the eye, the ear, etc., plays an important role.
X. Street performance demands crowd calling. The crowd is dynamic and the way of calling should be unique with a specific variety and style to draw the audience's attention.
XI. As streets are open, creating special effects in performances through lighting is possible only when dark. Sound, coloured props and coloured dresses could be used for the same.
XII. Any performance in the open is vulnerable to chaos and nuisance. The herd mentality of the crowd needs to be controlled through designing of the performance so that the audience is glued to them.
XIII. Performance on streets is vulnerable to noise. Thus, space dedicated for different performances should ensure that they are sequenced properly to ensure that apart from the background noise the ones coming from different performances should not interfere with one another.

4.15 PRACTICE SESSION

i. A script writer for a cinema seeks human factors advice on ways and means of using text for casting of actors, directors and all those involved in the movie. She also needs to give some warnings for some violence scenes as well as warning against smoking and drinking alcohol.

Human Factors Direction

A. Upper- and lowercase should be used for casting of actors and other people. Lead role could be in all capital letters to emphasise.
B. The word "warning" to be in all capital followed by the type of warning in upper- and lowercase.

C. The movement of the texts for different names ideally should be from bottom to top or from left to right.
D. The warning text should be static and supplemented by a voice over reading the text.

ii. A script for a theatre is lacking emotion after the first show. The director of the theatre asks for your expertise in incorporating human factors elements to enhance emotion in the script. The theatre is all about the story of a poor boy who struggles hard to get a high paying salary in a corporate sector.

Human Factors Direction
A. Introduce popular songs which were once very hit in the market.
B. Use lighting along with music to create emotion in specific scenes.
C. Advise the director for asking the actors to modulate their voice in a high pitch while delivering emotional dialogues, accompanied by music, for example, violin.
D. If needed voice-overs in emotional scenes in theatres could augment the emotional element.

iii. A magician performing on the street with audience all around seeks advice on human factors tips for entertaining the audience.

Human Factors Direction
A. Use brightly coloured props
B. Use bright orange coloured dress to draw the audience's attention.
C. Loud music or dance performance is recommended.

iv. A street magician is having a tough time in getting crowd for his performance. He seeks your human factor intervention for increasing the crowd flow.

Human Factors Direction
A. The location of the performance should be at the place where the crowd flows backing home from office. The crowd is tired and possibly some won't mind a performance.
B. Some small sample magic could be shown while moving on the street among the crowd so that they are interested in watching the main magic show.
C. Loud jingle music can help in attracting the crowd's attention.
D. Bright coloured dresses attract people and would help in attracting crowd.
E. Some dynamicity in the form of a big balloon tied to the ground, a person on stilt walking around, or some dance performances can help in drawing the crowd's attention.

F. Distributing performance-related information in the form of mask (so that it can be used), cap can also help in spreading the information related to the performance.

v. A band is performing on the roadside. There are some eight members of the band. They noticed that of late the crowd that assembles becomes chaotic. The group is asking for human factors advice in controlling the rouge crowd.

Human Factors Direction

A. Try organising the crowd through temporary barricades which makes the crowd stand in an organised fashion.
B. Give a preview of the performance before your start.
C. Make sure that the performers move around a specific zone only and not go beyond that.
D. Take requests from the crowd as to their favourite music they would like to hear.
E. Learn using humour in between the performance and talk to the crowd with regard to their feel of the performance.

Using Film and Video for Advertising

5

5.1 OVERVIEW

The human factors aspects of designing the storyline in tandem with the user needs and wants as well as for conveying specific messages of social relevance with specific emphasis on advertising are discussed here. The chapter discusses the human factors application in making of advertisements in digital medium, which leaves a long-lasting impression on the audience. Framing of dialogues along with their delivery from human factors perspective is also discussed. The human factors of "punchline" in the media (both film and advertisement) are talked about. Lastly, the chapter deals with the human factors aspects of ambience creation, connecting users with the product being advertised, and also how to design human factors aspects for eliciting customer emotion and how to measure the same are discussed here.

5.2 STORYLINE AND RECALL

I still remember a childhood advertisement that I had seen. Two people inside an aircraft are sweating. Suddenly the aircraft door opens and the two boys tie a rope to their legs, with the other end of the rope being fixed to the aircraft. A cold drink bottle is handed over to them. They hold the bottle with the cold drink spilling out of it and in that position both of them jump from the aircraft. They start hanging upside down with the rope tied to their legs and in that position keep on drinking from the bottle and the name of the soft drink pops up on the screen. The advertisement and the storyline contained characters and

DOI:10.1201/9781003506409-5

situations which are from the real world, but coupled with some weird acts because of which I still remember it. Only when there is something special in the storyline that you remember it for long as it creates an emotional "landmark" in your life. It's just like you remember any physical landmark in space.

The biggest challenge for media professionals is how to create a proper storyline for their movies and advertisement films and at the same time ensure that people are able to remember them (recall). The problem here is that your audience is bombarded with lots of information all around. Among these oceans of information, you have to draw the attention of your audience, and at the same time ensure that they have a good recall of the movie or the advertisement. This is important because it is on this approach that the business depends.

Duggi and Leto are two kids who watch movies a lot. Their parents ensure that they should watch good movies which have some good lessons in them and do not comprise violence and action. As they sit for watching the movie on the television, they start browsing through different channels. Finally they came across a channel where a classical movie was scheduled. There was an advertisement going on and they had no option but to watch the advertisement. The advertisement was related to a soft drink. A boy was drinking the soft drink sitting on the tree top. Suddenly he jumps from the tree but lands up right in front of an elderly couple walking below the tree. The couple is surprised but the boy with the cold drink bottle still in his hand offers the remaining drink to the couple. The lady grabs the bottle and finishes the same at one go with a sense of pleasure. On the screen the name of the soft drink is flashed with a jingle. Duggi and Leto are amused and become curious about the drink.

The above is an example of a storyline built for creating interest and for easy recall. The story line was funny and weird. In normal circumstances this does not happen. Offering the unfinished soft drink to the couple and the lady accepting that was funny and was targeted at the young population. No one watches an advertisement with interest, but here the content made it interesting. Certain scenes increased their recall value and users would recall them after a long time. These were the boy jumping from the tree, the lady finishing the drink at one go, etc. This is an example of a proper storyline and which had a high recall. After this the film started and the young people started watching the same but the advertisement left a lasting impression in their mind. These types of storyline which are a combination of what we see daily (that generates a pattern in the brain) and something which is not as per expectation, leave a mark in the memory. This mark becomes the point of reference in the brain with respect to which retrieval of information becomes easy. Later on when the audience shift their focus on something else, they still remember that "something different" that they saw. Later on when they look at the product it's easy for them to recollect the details because the product has by then become an element of the small advertising movie.

Mr Nitai and his wife Ms Rekha go for a movie which is based on the life of a very popular soccer player Mr Leto and his struggle in becoming a star player of the country "dhichkao". The movie started with the soccer match in which Leto played very well. Leto scored a hat-trick and that led to the victory of his country. From here the movie starts with the casting and showcasing the childhood of Leto and his struggle in becoming a successful soccer player. The couple watched the movie with rapt attention as they could relate the struggle of their own son who is a very successful business man in the fishery sector of the country where they belong to. So the storylines should align with the life of the common person. The moment this is done people are able to relate them with their daily lives and they get more engaged with the move and the story. The movie ends with the advertisement of a health drink that Leto drinks. This impresses Rekha and Nitai as they see their son in Leto. The storyline matches the audience's aspiration!!

5.3 KICK POINT IN ADVERTISEMENT

Kanu is walking along the busy streets of Beltala. This is his first trip to the place and he wants to buy something unique. As he looked around he found banners all around, advertising different products. He is totally confused as to what he should buy and what not. Finally he decides not to buy anything and keeps on walking and enjoying the place. In any advertisement or commercial (both static and dynamic) getting the customer attention is a big challenge. Static advertisements are the ones that are available as prints and found in magazines, newspapers and on the roads. If you just display your product or service and print it, there are chances that it would get hidden among thousands of such advertisements around. Advertisements which appear as movie clips and are shown between television shows, in movies or played in public domain also face similar challenges.

The biggest problem is that your users are "bombarded" with information from all around. In the advertisement world these information relate to various products and services. It becomes a problem for the user to look for the advertisement that would be relevant for them. For example, let's imagine that you are new to a town. You are looking for some good shoes as your shoes have all of a sudden torn off completely. You are desperately looking for some shoe shops but unable to find one. You see advertisements all around for different food products like ice cream, fruit juice and jumpers but could not locate a single advertisement for shoes. It was a sale season for different products and you were expecting that shoes would be on sale as well. You ask a person on the street and he points to a small banner almost hidden behind a big banner displaying sale on some good sneakers along with the name and address of the shoe store. The shoe store banner was so small that it was hidden behind the bigger banner and not visible. This is the problem with advertisement. They need to "pop" up from the surrounding and "make their presence felt". How do you ensure that?

To address this either you choose a place so that people can see it or you think of using a different form. A giant cut-out of a shoe (Figure 5.1) in the

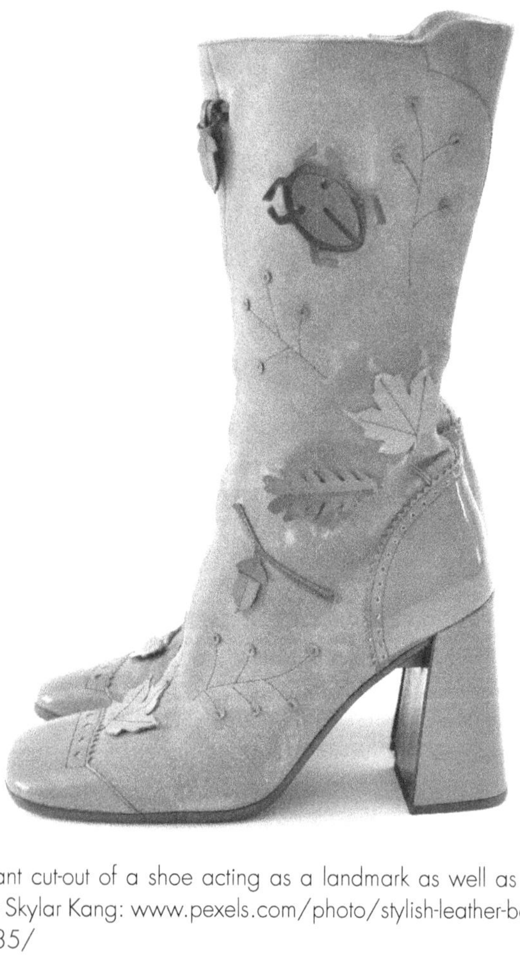

FIGURE 5.1 Giant cut-out of a shoe acting as a landmark as well as directional signage. Photo by Skylar Kang: www.pexels.com/photo/stylish-leather-boots-in-modern-studio-6046235/

middle of the street would be a nice surprise element for the pedestrians, with the toe of the shoe pointing towards the store for helping people navigate in that direction. For example, if the majority of advertisement banners are rectangular, yours one could be circular or it could be displayed on a balloon tied to the ground. The next would be to use certain language and visuals that evoke people's emotion. For example, you could display the photograph of a favourite film hero wearing a nice pair of shoes and a small squirrel sitting on the black shoes. Along with this there could be some small white flowers lying close to the black leather shoes. This composition in the advertisement creates an environment which generates a storyline with a bit of emotional touch with a little innocent animal, the squirrel, and some tiny white flowers. You may call this the "kick" in the advertisement for conveying information just like an alcoholic gets a kick while drinking alcohol. Such "kick" leaves a long-lasting emotional mark on the users and they remember it for long and it also drives them to buy the product.

If the advertisement is dynamic, then you are in a better position as dynamic information is better registered in the brain. Any good advertisement clip with a strong storyline from the real world leaves a mark (emotional mark) on the customers and they retrieve the product long after they have seen the same. For example, a very old advertisement for a health tonic shows a grandfather and the grandson. The grandson approaches the grandfather to play a game of badminton. The grandfather says he would play but before that both of them need to have a teaspoon of the health tonic. So they both have a teaspoon of the health tonic and they start playing badminton. Customers remember the health tonic because of the kick point in the advertisement which was quite emotional in terms of the sweet relationship between the two generations, the grandson and the grandfather. In any close-knit society these relationships are very emotional and people value them a lot. So any advertisements around these have a huge impact on the customers' mind and it also increases the recall value of the same.

5.4 DIALOGUE DESIGN AND DELIVERY

The framing of dialogues in film, video and theatre also creates patterns that are auditory in nature. These auditory patterns in dialogues are registered easily and with minimal error. The audience also remember these dialogues with patterns for a long time. The famous dialogue of James Bond, "My Name is Bond, James Bond", is a dialogue that people remember for generations. It is because of the pattern of the dialogue along with the voice modulation of the actor that etches a long-lasting impression in the minds of the audience. Thus, while writing scripts it's important that dialogues are "designed" in a manner that there are variations and one has to ensure its proper delivery ensuring that the dialogue is not monotonous and there is modulation of the tone of the actor delivering the same. As people are bad in short-term memory it's better to give adequate pause in between words and ensure that dialogues are short and articulated with adequate pause in between. These types of small dialogues with punch in them when linked to a product do a very strong advertisement of the same to the audience. It has been seen that in advertising any product, if the benefits of the products are repeated in different ways, through textual information, movies, model of the product, etc., it sells well and people tend to accept them.

5.5 THE PUNCHLINE

The punchline in any performing arts plays an important role in impressing the audience. That's the reason why some dialogues are remembered by the audience even years after the performance. These again leave behind emotional landmarks in the minds of the audience with respect to which the retrieval of information becomes easy. To develop these punchlines in dialogue one has to get to know the pulse of the audience. It should be then designed accordingly. The liking of the audience could be along the lines of violence, romance, emotion or sadness. Thus, the dialogue punchline should follow similar patterns. Just like in boxing, the final punch with right impact and location at times decides the fate of the game and the opponent falls down and is unable to get up and the person who punched is declared the winner. Similarly, in film, video and theatre advertisements, selecting the right punch word is important. These words should be selected in tandem with the expectation or visualisation of the common people. For example, if you are planning to launch a new ice cream in a town specially meant for diabetics, then your focus should be to highlight the aspects of the product being "sugar free", "no artificial sweetener added" (as diabetics prefer not to take them as they cause side effects), "zero calorie", etc. These are the words which diabetics are always concerned with and are the ones which can form the punchline for the advertisement for making it more reachable to the target customers through the advertisements. As these are engraved in the brains of these target population it's better to use them for advertising the product. Similarly when a movie for advertising a product or service is made, the punchline should be picked up from some major area of interest of the population. It could be "corruption-free politics" which people aspire to see. This being a dominant topic is again etched in people's mind and they would love to see them in the movie script and thus it becomes the central theme or punchline for the movie in accordance with the target audience's needs and wants. This could then be linked to a new newspaper being launched which targets a corruption-free society.

5.6 CREATING THE AMBIENCE AND CONNECTING THE PRODUCT

An advertisement is being aired on television on a new township on the bank of a river "Shoshti". It's an apartment for the elderly. The advertisement starts with a walkthrough through the lush green campus and the river side. Then it takes the audience to the reception area and finally it leads to a sample apartment completely furnished and with a beautiful view of the setting sun through the window on the river "Shosti". A careful look at the advertisement reveals how the ambience of the accommodation was created. First an overview of the township was given, giving users an idea about the location of the township. Then the exterior was shown as if the customers are entering the township, touching upon each and every part, and then finally they were taken inside the apartment. To make the ambience better the sunset was shown through the window on the river. This storyline was developed exactly in tandem with the "way of thinking" of retired and elderly population whose daily lives are slow and go on in this way. It starts with a leisurely stroll around the locality (here the campus, green lawn and river side) and then they enter their respective apartments. After sunset the elderly feel safe to be back home. So the sunset through the window is a perfect depiction of that aspiration of the users. This is how the way of thinking of users should be banked upon.

A new product is being launched. It helps user to measure their level of hunger. They have named it as "hunger-meter". it is a useful product (Figure 5.2) for the mothers for monitoring the hunger level of their newborns as they are not able to speak. The company creates an advertisement for the product. The film depicts a child sleeping and the mother is looking at the child and wondering if she would wake the child and feed her, or the child is already full. The caption in the advertisement says "no need to worry", just touch the hunger meter to the infant's body and it will indicate whether the infant is hungry or not. Again an attempt to connect the product to the user. The mother is always worried about the diet of the child and especially new mothers have no experience as to when to feed the child and when not to. This aspect has been picked up to portray the product and thus the "way of thinking" of the

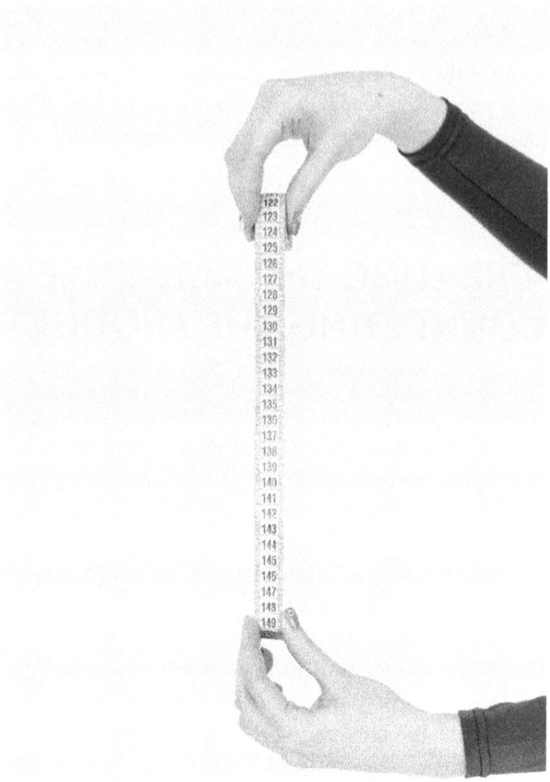

FIGURE 5.2 The hunger meter advertisement. Photo by SHVETS production: www.pexels.com/photo/woman-showing-measuring-tape-for-checking-body-forms-6975471/

mother is aligned with the product and the same is advertised so as to connect well with the mother and also factoring in the ambience of the bedroom and the sleeping child.

5.7 DESIGN AND MEASURING CUSTOMER'S EMOTION

In advertisement playing with the customers' emotion is the most important element. All scripts and sequences are to be designed keeping in mind this emotional aspect. Always remember that emotion is very difficult to measure in humans. Your aim should be to create that framework in your advertisement that evokes that specific emotion that you are looking forward to. If chocolate evokes people's emotion irrespective of their age, then an advertisement for a new type of chocolate should try to evoke customer emotion. One of the ways you can get an insight into people's emotion is to ask them what makes them emotional or nostalgic. Once you get this data you can design your advertisement script accordingly.

Leo is a Labrador dog of a young boy called "Hari". Hari wants to treat Leo with special "dog's chocolate", as the vet has asked him not to give Leo chocolates meant for humans. Every time Hari takes chocolate he feels guilty because he cannot share the same with his best buddy Leo. At this juncture a company comes up with a new advertisement for "dog's chocolate". The script starts with a young boy taking his favourite dog to the nearest park. They play there for some time. Then both of them get tired and decide to rest under a tree. The boy takes out a chocolate and was about to eat when he finds his favourite dog staring at him (Figure 5.3). He feels bad and keeps the chocolate back in his pocket. An elderly person who was watching the boy and the dog comes forward. The elderly person gives the boy a small chocolate and says "this is for your best friend", don't worry it's a "dog's chocolate". The boy looks at the elderly person with surprise and takes the chocolate from him. He offers the chocolate to the dog and takes out his own from his pocket. Both of

FIGURE 5.3 Leo and the Petcolate. Photo by Blue Bird: www.pexels.com/photo/woman-with-labrador-on-grassy-lawn-7210714/

them start enjoying their respective chocolates and thank the elderly person. The name of the dog's chocolate flashes on the screen "Petcolate" for your favourite doggy!! This is where the advertisement plays with the customer's emotion as they have measured the emotional level of the pet and humans. They are able to create the framework through the storyline which is a familiar storyline in any household and hence people are able to relate to it very easily and get emotional. The retrieval of information about the chocolate becomes easy because the storyline creates an emotional landmark in one's mind and we know the retrieval of information becomes easy with reference to any landmark, be it physical or emotional.

5.8 KEY POINTS

I. Create a storyline which is short and easy to recall.
II. Make a point in the advertisement a little different from the norm. That becomes the kick point which people remember.
III. Your dialogues need to be brief, and delivered with voice modulation to avoid it to sound monotonous.
IV. The punchline is the identity of the product being advertised. Make sure that it's the climax in the advertisement.
V. For creating ambience try to take the user to the context and film accordingly. Products should depict their function in real-life situation.
VI. Create the framework for eliciting emotion by studying people's daily life and the context and relations that could evoke emotion.

5.9 PRACTICE SESSION

i. A company launching a special pen for college-going students is looking for your human factors intervention in designing their script for an advertisement. The advertisement should make an impact on the college students so that they are encouraged to buy the pen.

Human Factors Directions
A. Look for a pattern in the daily life of the college student.
B. Based upon the pattern script a movie.

C. Focus on the pen as a style statement and a product of utility.
D. Have young boys and girls act in the film.
E. Show the college environment.
F. Give a dash of romance to the film.

ii. A company is looking for your expertise in enriching an advertising script for a natural fruit juice meant especially for people suffering from hypertension.

Human Factors Directions
A. Mention the ingredients in the drink..
B. Show how the drink is made naturally to build trust among people.
C. If possible include a small clip of a renowned doctor who is recommending the drink for hypertensive people.
D. Show that people after having the drink become energetic.
E. Do not forget to show the limit for the drink per day.

iii. A new music school for senior citizens is coming up in town. The company which is making the advertising film for the same has requested you for your human factors intervention in designing the film including the script. What would you advise them?

Human Factors Directions
A. Highlight that life starts after one retires.
B. Talk about the need to remain active and healthy.
C. Show the importance of music as a therapy for better sleep, mood and overall health.
D. The kick point should be that it's never too late to pick up music as a full-time hobby.

6 Exercises on Human Factors in Performing Arts

Exercise 1

A magician is coming to the town hall to perform. He needs some human factors advice to ensure a good show and to ensure that the audience is happy. Your expertise has been solicited to ensure a seamless performance.

Human Factors Direction

A magical performance is a combination of tricks, illumination, music and acting. First ensure that the space requirement on the stage matches with the demand of the show. Check and calculate the individual workspace envelope and then calculate the total space requirement for the required number of people on stage. A very crude way of calculating the individual workspace envelope on stage is to ask the performers to stand on the stage and make them fold their forearm and make the fingers of the palm touch each other near the chest. An imaginary circle drawn taking the elbow-to-elbow distance as the diameter gives a rough estimate of the space requirement for a performer. Likewise calculate whether the stage can accommodate all the performers in one scene. Add some extra allowance for props, movements, etc.

From the stage and from the audience seating ensure that the audience is able to see all the parts of the stage when seated. If this is not taken care of, they would not enjoy the magic show.

All lighting system should preferably fall on the performers on the stage. Lighting should not fall on the audience's eyes, else it would lead to disturbance for the audience. Musicians should be positioned a little below the stage so as to ensure that they are visible but should not come in the way of the performance and the audience.

To ensure that the tricks are not detected by the audience full control of the light and the music system should lie with the magical team. In case some tricks get exposed and to distract the audience the stage light could be dimmed

DOI:10.1201/9781003506409-6

and the music intensity could be increased proportionately to create a different ambience.

Ask the magician if this is his first show, preferably not to repeat a trick for the second time even if requested by the audience. Tricks have a certain pattern and the human eyes are good at detecting patterns if it's repeated. So, there might be some in the audience who might detect the pattern and the trick.

Exercise 2

There is a circus in the town. They have just started constructing the tent at the venue. The circus team is seeking some human factors advice for ensuring a fantastic performance and satisfying the audience. Your help as an expert in the field is solicited.

Human Factors Direction

Like example 1 calculate the dimensions required by the performers and depending on that the ring size should be decided. Ensure that the ring is visible from all the seats. As the circus contains certain shows in the upward direction near the top part of the tent (the trapeze) it requires the audience to look upwards. For those sitting close to the ring might have to strain their necks to tilt their heads backwards which is painful. So the first few rows of seats in proximity to the ring could be tilted a little backwards to ensure proper visibility of the trapeze acts.

Each event should be accompanied by relevant music and if possible popular music should be played so that the majority of the young audience could relate and enjoy the same. Any show if it becomes monotonous could be interspersed by shows by the circus jokers, which would break the monotony. Before the daredevil acts the end results of the same should not be disclosed, else the surprise element is lost.

It's very important that the main event of the circus is advertised among the public in the town and most importantly the duration of the circus should be clearly mentioned. This helps the audience willing to watch the circus take prudent decisions like the preferred show timing, which day of the week to go, etc.

Exercise 3

A new director is making a film on the life of the unemployed youths in the country. He seeks human factors advice starting from shooting the film to editing. What advice would you give him so that the film gets the maximum appreciation?

Human Factors Direction

In the shooting of the film, the director has to ensure that the actors are not fatigued and given adequate rest in between working spells. Fatigue leads to performance degradation and the same is reflected in the look of the actors, which on camera could look weird. While shooting outdoors and in extreme hot and humid conditions care should be taken to adequately hydrate all actors. They should also rest in cooler areas under shade at regular intervals to delay the onset of fatigue.

To prevent distraction from the public around while shooting in public places it's better to explain to the public in brief about the shooting so that their level of curiosity is addressed to a certain extent. Those scenes which require maximum action and less of dialogue delivery should be shot in public as they are least affected by public distraction.

While scripting try incorporating the emotions of the unemployed, parent's agony, poverty, etc.

While editing the film care should be taken to use upper- and lowercase letters for presenting majority of the information as that enhances better readability. Important casts and notices, "no smoking", etc., should be in all capital letters to indicate their relative importance.

The use of pure red and green colour should be avoided as a large proportion of the world's male population are pure red and green colour-blind and would perceive them in shades of grey.

Exercise 4

An advertising company is looking for a film for their upcoming fairness cream for males who have a dark complexion. They have their in-house script writers and film makers but are looking for specific human factors inputs for increasing the impact of the film on the probable customers. What would be your advice to the film-making team?

Human Factors Direction

The public needs to be give a clear perception that fairness creams are not necessarily meant for making the skin look lighter. Fairness means good-looking skin and that can happen in those having a dark complexion. This mental model of the users needs to be developed through the advertisement. Possibly a celebrity who is having a dark complexion could be sighted as an example and shown that he uses the fairness cream to achieve what he is today in terms of his skin fairness.

The tag line should be as follows: fairness for dark skin which ensures its health, shine and attractiveness. With respect to this people would be able to retrieve the information and remember the product.

The next thing that people would like to see are the ingredients and if they are natural. The mental model of when one buys a product like this is as follows: does it contain natural ingredients, how long it should be used, the amount and the cost. If these are highlighted along with the main features of the cream, it should elicit curiosity among the would-be customers of the product.

Exercise 5

A band wishes to perform inside the underground metro station after getting requisite permission from the competent authority. The band comprises eight members and they decide to perform from morning till evening inside the station. As they are new, they seek human factors advice as to where and how should they arrange their setup for making maximum impact on the passengers travelling by this mode.

Human Factors Direction

There are a couple of aspects when it comes to performing in a setup which is not normal. First, locate a place to be in the direct visual cone of the commuters without obstructing their movement. Corners and roundabouts are better options and the wall could be used as a backdrop for performance. After calculating the space requirement for each performer and their equipment (Exercise 1), a line with a tape could be marked on the floor, demarcating the performance area. Make sure that the audio system is located only in the territory of the band and not spread all around as that might infringe with the announcements made on public address system. The band need not perform continuously but only during rush hours and when there are crowds during festive seasons, holidays, etc.

It's better to perform bands of different music genre. This would ensure that it caters to all types of audiences and as the audience are listening in a different setting while standing, they would also not get bored.

To make the band lively the band might invite people from the audience who would like to play along with them. This would facilitate interaction with the audience and make the performance interactive. People who were into music would always love to showcase their talents in public domain!!The audience could be invited to sing with them.

Exercise 6

A newly opened bar has arranged for a vocalist to perform along with her musicians in the evening in front of the guests. It's a small bar and they are seeking human factors advice as to how to use this performance for increasing the customer footfall inside the restaurant.

Human Factors Direction

Performance in a bar is one of the major attractions for increasing customer footfall. First, locate a proper space and calculate the area required to accommodate the requisite number of people as shown in the previous exercises. Make sure that the performance is done on a raised platform or stage so that it's within the visual cone of all customers.

There should be an advertisement in front of the bar about the performance so that people can take a decision of entering the bar for its added attraction.

The use of special lighting on the performers would highlight them in the dark ambience of the bar. Make sure that the location of the stage is in an "expected area" inside the space. When people enter a space, they expect certain things in certain areas. For example, the reception should be when you enter, the bar counter very close to it. Similarly, the stage for performance should be in the proximity of the bar counter so that people do not have to look around for it.

Make sure the audio volume is low so as not to interfere with people's conversation in the bar. The illumination should be dim so as to maintain privacy as well a setting the mood of the bar for drinking.

Exercise 7

A musical night has been organised on the playground of an undergraduate college in the town. An event management company is organising the event and they are setting up the stage for the same. The stage has to be set up on the playground, with the audience having no specific seating arrangement;, they could stand, sit on the grass or bring in their own foldable chairs if they want. Your human factors expertise is solicited in making the event successful. Remember the crowd is mainly the young college students.

Human Factors Direction

First, the stage dimension has to be calculated as before depending on the number of performers. Next, the height of the stage should be such that it's within the visual cone of the audience while sitting and standing both. Thus, it might be a little higher than a traditional venue where everyone is seated. The allowance in this case for the stage size should be more as more movement is expected, it being a college programme.

In a venue like this crowd control is very important, otherwise cases of stampede might happen. The venue should be fenced all around so that the crowd cannot trickle from anywhere but only through defined entry routes. There should be four to six exit routes to ensure rapid and uniform crowd dispersion after the show. The crowd should be made to sit or stand in specific zones and only when one zone fills up should the crowd be allowed to move into the next zone. This would ensure that the crowd is uniformly distributed so that when it comes to emergency evacuation there are separate exits for each zone and no bottleneck.

As it's a performance in open the public address system should be uniformly distributed all over the venue and not confined to one place. At the beginning of the show the audience should be briefed about the emergency exit routes as well as any necessary protocols to be followed so that they have a mental model of the space and what to do and what not to.

Exercise 8

A street play has been organised by a group for raising awareness against COVID-19 pandemic among the citizens. The organisers want to stage this at a very busy street near the milk market in the town. Your expertise as a human factors specialist has been solicited to ensure the best performance by the group and maximum impact on the audience.

Human Factors Direction

The space area needs to be worked upon like the previous exercises and based on this the area needs to be demarcated for the performance. The location has to be chosen such that it falls within the people's normal path to and from the milk market so that they cannot miss it. The performers if they are presenting in bright daylight should preferably wear bright coloured clothes so that they are visible from a distance as there would be public all around.

The dialogues should be crafted in a manner so that they are short, have familiar words that people understand and should be delivered at a high pitch so as to be audible in the noisy environment. Coloured props should be included which would attract the audience's attention and especially if they are of different shapes and odd sizes which is not found in the environment around. For example, a big balloon, a huge cardboard cube sor a very big six-foot high stethoscope or a giant mask, etc.

Exercise 9

A new film is being made. The theme of the film is to save the animals and try to increase their number. Your advice as a human factors expert has been solicited to help in making the story, script robust and lively and also help in the editing of the film.

Human Factors Direction

The storyline should have a flow through characters with which people could relate. Have children in the film as they are the like animals and both are equally innocent. Children in the film would also showcase the future of the planet once they grow up. Have different animals starting from domestic to the wild ones and show all of them in their habitat and as good friends of humans. This would create a positive mental model in the audience and help in developing a love for animals and also they would feel the need for their existence on this planet.

At the editing stage animal voices, sounds of jungle and music could be used to create that natural ambience in the movie. Some amount of romance between humans and animals would add an extra layer of emotion to the film and make it acceptable. Have the captions for castings, etc., in colours which are taken from the forests and animals, which would create a mental model and enhance people's perception of the importance of wild life conservation.

Exercise 10

A new theatre company is planning to stage its first performance in the town. They need help in writing the story, script writing and delivery of the same. They are looking for some human factors inputs in their endeavour. The company wants to target the college students who are into different types of addiction and finally spoil their career in the long run.

Human Factors Direction

Humans love to imagine, dream and visualise. Theatre as a medium can help them do all these through performances. Choose the storyline and the script so that it portrays the need for an addiction-free life and how it helps people stay healthy. Show the ill effects of addiction and how it makes people drift from a beautiful life and lead a life full of suffering. No human would like to see sufferings and once shown would possibly understand the ill effects of addiction. Some career paths might be depicted and how they lead to a very satisfying lifestyle. This would create a mental picture among the audience with regard to their choices in life and based on this they would be in a better position to take a prudent decision as to what they would like to do with their precious life.

Human Factors Application in Performing Arts: Some General Directions

A. First identify who are the performers and the exact composition of the audience.
B. Have an estimate of the number of performers and the audience.
C. Calculate the space requirement for the performers and then calculate the same for the audience based on their body dimensions and factor in the movement dynamics.
D. Make sure there is an eye-to-eye connection between the performers and every audience.
E. The performers should be audible to every audience, especially to those audience seated at the rear.

F. Ensure proper crowd control by providing large number of entries and exits both at the four corners of the venue and at the centre of the venue.
G. Scripting for stage shows should ensure specific "kick points" taken from real life which the audience would be able to recall when the performance is over.
H. The attention span of the audience is very short especially when the performance takes place in informal settings. Here focus on the costume, short dialogues and overall shorter duration of performances.
I. A successful performance should always factor in audience reaction and thus moving close to them or even performing very close to them for a few minutes makes the performance much more entertaining.

Bibliography

Alderson, J., Hopper, L., Elliott, B., & Ackland, T. (2009). Risk factors for lower back injury in male dancers performing ballet lifts. *Journal of Dance Medicine & Science*, 13(3), 83–89.

Berry, N. M., Prugh, W., Lunkins, C., Vega, J., Landry, R., & Garciga, L. (2015). Selecting video analytics using cognitive ergonomics: A case study for operational experimentation. *Procedia Manufacturing*, 3, 5245–5252.

Cabral, A., & Figueiredo, C. (2014). Costume design: Ergonomics in performance art. *Advances in Ergonomics in Design, Usability & Special Populations: Part*, 1, 511–520.

Colucci, L. A., & Klein, D. E. (2008). Development of an innovative pointe shoe. *Ergonomics in Design*, 16(3), 6–12.

Dommerholt, J. (2010). Performing arts medicine–instrumentalist musicians, Part II–examination. *Journal of Bodywork and Movement Therapies*, 14(1), 65–72.

Du Yihang, P. Y., Ke, N., Yao, Z., & Hui, X. (2023). Movement description method and application for dance of Chinese traditional drama. *Digital Human Modelling and Applied Optimization*, 41–50.

Garcia, M. C. (2013). Experimenting with the circus amongst physical education teachers. *Staps: Revue Internationale des Sciences du Sport et de l'Éducation Physique*, 34(102 2013/4), 47–60.

Hancock, P. A. (Ed.). (1999). *Human performance and ergonomics: Perceptual and cognitive principles*. Elsevier.

Li, D., & van der Veer, G. C. (2019). Cognitive ergonomics on the move: Art ecosystems are a new application domain. In D. M. Murray, G. C. van der Veer, G. de Haan & A. Dittmar (Eds.), *ReCogErg@ ECCC* (pp. 12–16). CEUR-WS.

Liang, P. (2019). Towards a probody aesthetics: Ageing and occupationally damaged bodies in performance. *Research in Drama Education: The Journal of Applied Theatre and Performance*, 24(2), 211–224.

Lund, A. M. (1996). Advertising human factors. *Ergonomics in Design*, 4(4), 5–11.

Millet, B., Chattah, J., & Ahn, S. (2021). Soundtrack design: The impact of music on visual attention and affective responses. *Applied Ergonomics*, 93, 103301.

Oreb, G., Ružić, L., Matković, B., Mišigoj-Duraković, M., Vlašić, J., & Ciliga, D. (2006). Physical fitness, menstrual cycle disorders and smoking habit in Croatian National Ballet and National Folk Dance Ensembles. *Collegium Antropologicum*, 30(2), 279–283.

Shorrock, S., & Williams, C. (Eds.). (2016). *Human factors and ergonomics in practice: Improving system performance and human well-being in the real world*. CRC Press.

Svanaes, D., & Seland, G. (2004, April). Putting the users center stage: Role playing and low-fi prototyping enable end users to design mobile systems. In *Proceedings of the SIGCHI conference on Human factors in computing systems* (pp. 479–486). ACM Digital Library.

Twitchett, E., Brodrick, A., Nevill, A. M., Koutedakis, Y., Angioi, M., & Wyon, M. (2010). Does physical fitness affect injury occurrence and time loss due to injury in elite vocational ballet students? *Journal of Dance Medicine & Science*, 14(1), 26–31.

Wiberg, V. (2012). Communication of ergonomics in building and construction. *Work*, 41(Supplement 1), 4111–4115.

Wilson, J. R. (2000). Fundamentals of ergonomics in theory and practice. *Applied Ergonomics*, 31(6), 557–567.

Index

Note: Page numbers in **bold** indicate figures on the corresponding pages.

For Product Safety Concerns and Information please contact our EU representative GPSR@taylorandfrancis.com
Taylor & Francis Verlag GmbH, Kaufingerstraße 24, 80331 München, Germany

www.ingramcontent.com/pod-product-compliance
Ingram Content Group UK Ltd.
Pitfield, Milton Keynes, MK11 3LW, UK
UKHW022157190926
475702UK00003B/75
9781032826912